Animal Entanglements

RADICAL CULTURAL STUDIES

Series Editors: Debra Benita Shaw and Angie Voela (Centre for Cultural Studies Research, University of East London, UK)

The Radical Cultural Studies series publishes monographs and edited collections to provide new and radical analyses of the culturopolitics, sociopolitics, aesthetics and ethics of contemporary cultures. The series is designed to stimulate debates across and within disciplines, foster new approaches to Cultural Studies and assess the radical potential of key ideas and theories.

Sewing, Fighting and Writing: Radical Practices in Work, Politics and Culture, Maria Tamboukou
Radical Space: Exploring Politics and Practice, edited by Debra Benita Shaw and Maggie Humm
Science Fiction, Fantasy and Politics: Transmedia World-Building Beyond Capitalism, Dan Hassler-Forest
EU, Europe Unfinished: Europe and the Balkans in a Time of Crisis, edited by Zlatan Krajina and Nebojša Blanuša
Postcolonial Interruptions, Unauthorised Modernities, Iain Chambers
Austerity as Public Mood: Social Anxieties and Social Struggles, Kirsten Forkert
Metamodernism: Historicity, Affect, Depth, edited by Robin van den Akker, Alison Gibbons and Timotheus Vermeulen
Affect and Social Media: Emotion, Mediation, Anxiety and Contagion, edited by Tony D. Sampson, Stephen Maddison and Darren Ellis
Gender, Sexuality, and Space Culture, Kat Deerfield
Work That Body: Male Bodies in Digital Culture, Jamie Hakim
Writing the Modern Family: Contemporary Literature, Motherhood and Neoliberal Culture, Roberta Garrett
Revisiting the Nomadic Subject: Women's Experiences of Travelling Under Conditions of Forced Displacement, Maria Tamboukou
Walking Networks: The Development of an Artistic Medium, Blake Morris
Animal Entanglements: Muddied Living in Dog–Human Worlds, Erika Cudworth

Animal Entanglements

Muddied Living in Dog–Human Worlds

Erika Cudworth

ROWMAN & LITTLEFIELD
Lanham • Boulder • New York • London

Published by Rowman & Littlefield
An imprint of The Rowman & Littlefield Publishing Group, Inc.
4501 Forbes Boulevard, Suite 200, Lanham, Maryland 20706
www.rowman.com

86-90 Paul Street, London EC2A 4NE

British Library Cataloguing in Publication Information Available

Library of Congress Cataloging-in-Publication Data Available

Names: Cudworth, Erika, 1966- author.
Title: Animal entanglements : muddied living in dog-human worlds / Erika Cudworth.
Description: Lanham : Rowman & Littlefield Publishing Group, [2024] | Series: Radical cultural studies | Includes bibliographical references and index.
Identifiers: LCCN 2023057662 (print) | LCCN 2023057663 (ebook) | ISBN 9781538180198 (cloth) | ISBN 9781538180211 (ebook)
Subjects: LCSH: Human-animal relationships. | Human-animal communication.
Classification: LCC QL85 .C8293 2024 (print) | LCC QL85 (ebook) | DDC 590--dc23/eng/20231218
LC record available at https://lccn.loc.gov/2023057662
LC ebook record available at https://lccn.loc.gov/2023057663

For Kevin (1996?–2012), hair shedder extraordinaire, full of vim; and Gypsy (2006–2021), beautiful creature of silent power. Also, for Gypsy's 'whimpet' kin, favourite brother Tlachito (2006–2022) and their good friend Rio (2010–2023), big boy, full of joy.

In memory, also, of Thumbelina (1995–2013), amiable buffoon with opposable thumbs, and Penny (2004?–2020), Queen of Mean, who got her kicks striking fear in others (cats, dogs, frogs, humans—all fair game).

Tony Cudworth (1936–2010) died while Social Lives with Other Animals was in press; so here, Dad, this one is for you, too. Life is much harder now that you are not on the end of the phone knowing the answer to pretty much everything.

Contents

Acknowledgements

This book has been a long time coming and much has happened since my colleagues in the social sciences at the University of East London granted me some sabbatical leave to wander about on Walthamstow, Tottenham and Hackney marshes in London's Lee Valley Park. This time enabled me to pay attention to what was happening on dog walks I had previously taken for granted and to talk to people about their lives with companion dogs. When I first began to talk about this work at academic conferences, some members of the audience felt that what was said and what had been found, reflected cosmopolitan London dwellers' perspectives. This prompted a second phase of data collection when my time was divided between London and rural Leicestershire. In the data from the rural field site, there is some interesting material but nothing, perhaps surprisingly, that differed substantially from the stories from the metropolis! I would like to thank all those people I encountered walking with dogs, who gave me their time and shared their stories. I am grateful to the University of East London, in times past, for supporting my work and to some wonderful colleagues who over many years have read my work, provoked my thinking and shared struggles and laughter. In particular, I need to thank Delia Langstone, Deborah Benita Shaw, Maja Korać, Maria Tamboukou and last, but by no means least, my 'academic husband', Steve Hobden. For talk, support, reading stuff and great company, I need also to thank critical animal people Matthew Cole, Kate Stewart and Richard J. White.

The chapters in this book have been developed from some work that has previously appeared; any material is reproduced with kind permission. Chapter 1, 'Research Pack', is extended and developed from a paper in a special section on 'A Sociology of Multispecies Relations?', titled 'Now, Where Were We?' The Highs and Lows of Hunting Data with a Research Pack', in the *Journal of Sociology* (2018, 54, no. 4: 488–503). Chapter 2, 'Walking Edgelands', is significantly extended and developed from 'Posthuman Community in the Edgelands', in *Society and Animals* (2017, 25, no.

4: 384–403). A greatly abbreviated version of chapter 3, 'Sharing Home', appeared as 'Muddied Living: Making Home with Dog Companions' in the *International Journal of Sociology and Social Policy* (2021, 41, nos. 3–4: 424–39) and some of that material is reproduced here. Chapter 4, 'Eating Animals', is a significantly extended and heavily revised and updated version of 'On Ambivalence and Resistance: Carnism and Diet in Multispecies Households' from *Critical Perspectives on Meat Culture*, edited by Annie Potts (2016, 222–42 (Leuven: Brill)). Chapter 6, 'Working and Caring', is extended, revised and developed from 'Labors of Love: Work, Labor and Care in Dog-Human Relations', in *Gender, Work and Organization* (2022, 29, no. 3: 830–44). This book brings together many of the findings of my 'dog project', about which I have been thinking, re-thinking and reflecting over many years.

Dogs have been part of my life for all of my life, but particularly since the age of six when a dog came to share my home. My first 'adult dogs', Kevin the Jack Russell terrier and Gypsy the lurcher, were my research assistants for the project on which this book is based. I would like to thank them for all they taught me in their very different ways. Every dog and every relationship is unique and irreplaceable. Kevin and Gypsy certainly supported this research, but their passing has made parts of this book challenging to write.

Mud is a strange substance: water and towels never seem to really remove it from a dog's paws, and once it hits the bedsheets, it transforms into sand. I live in a home of mud and hair, but muddied living is more than this. It is about intimacy, relationality and messy everyday practices. Tinkerbell and Kali, I thank you for putting up with being removed from the desk/pile of paper/keyboard/lap so that I can write. Ruby, thank you for putting up with being ignored while I work despite your best attention-grabbing efforts, and for the walks which were later than I planned or earlier than you may have liked. Thank you, Vinnie, for the many hours of sleep therapy you have put into this book while keeping me company while it was written. Thanks, Dave, for cooking great food and for shedding less hair.

List of Figures

Introduction

Domination and Affection in Dogland

Nonhuman 'animals' have raised some interesting questions for what it means to be human, and for our consideration of the boundaries of the social world. Work on domesticate species has illustrated the extent to which we have entangled histories resulting in specific social formations. For Donna Haraway, important in understanding such co-constitution of the human with other animal species is embodied experience where we 'meet' as 'beings-in-encounter in the house, lab, field, zoo, park, office, prison, ocean, stadium, barn or factory' (2008, 5). In this book, I will be considering everyday lives with dog companions as they cohere around experiences of work, care, walking and the use of public space, food and eating, sharing the space of home, emotional bonds and relations of kinship. In looking at these aspects of the lives of dog and human companions, this book also frames the human-dog relationship in terms of matrices of power; understanding 'species' as socially constituted and as implicated in systemic relations of social domination.

Animals in the home kept as companions or 'pets' differ historically and geographically in terms of the species kept and kinds of relations humans have with them (Cudworth 2011, 144–47). In the West, human relations with 'pets' are the closest and most humanised across species, and as Jeffrey Masson rightly notes, dogs are the most domesticated of animals—that is, apart from humans (1998, 187). The domesticated dog (*Canis lupus familiaris*) has a huge global population—estimated at between 700 million and a billion animals (Smith et al. 2022, 1). Most of these creatures will be free roaming, that is, have unrestricted movement, whether animals are owned by individuals or by communities (such as 'village dogs') or are feral (surviving independently of human care). This book certainly does not present a picture of the lives of the majority of dogs, nor does it reflect the conditions in which most dogs live. As this book draws on a United Kingdom study, the dogs discussed are those who are kept in homes and are therefore 'owned', whose owners have a legal duty of (minimal) care. They are dogs who must legally be tagged and micro-chipped for identification purposes and can be reported and/or taken to the pound if found wandering without an owner in a public space (Rosling 2022). They are those dogs who, however well treated and much loved, are essentially, 'captive', animals that comprise only 15 to 20 per cent of the dog population globally (Pierce and Bekoff 2021, 26–28).

While such dogs might be a minority, the numbers of animals kept as companions in homes is very significant. In the United Kingdom, it is estimated that 62 per cent of households have pets (Statista 2022a); and the dog is the most popular, with 12.5 million dogs in 33 per cent of households (Pet Food Manufacturers Association (PFMA), 2022; Statista 2022b). Figures for dog ownership are higher in some European countries (Poland 43 per cent, Portugal 39 per cent, Romania 45 per cent) and the Americas (Argentina 66 per cent, Brazil 44 per cent) (Statista 2022c, 2022d). Across Europe, there are an estimated 93 million dogs kept as companions (FEDIAF 2022, 4). This is associated with significant levels of household expenditure. In 2018, annual sales of 'pet food' was 21 billion Euros, with pet-related products and services accounting for a further 18.5 billion (FEDIAF 2019, 5). The number of animal companions kept in UK households has risen year-on-year for the last decade and increased by 18 per cent since 2018–2019 (Statista, 2022a). Everyday lives, then, are often likely to be multispecies lives, and this is a cultural phenomenon worthy of scholarly attention. Sociologist and dog-lover Clinton Sanders (1999, 2) points out that the 'conventional inattention' to human relationships with animal companions has left 'a considerable gap in the sociology of everyday life'; even though mundane practices and everyday relations are 'at the centre of human existence, the essence of who we are' (Pink 2012, 143). This is the focus of this book: the everyday lived experience of people and—at least to some extent—the dogs with whom they share their lives.

Key reasons given by UK dog owners for living with dogs include that "they make me happy" (51 per cent), "for love and affection" (47 per cent) and companionship (35 per cent) (Statista 2021). As we will see over the course of this book, their presence questions human-exclusive notions of family and kin (Charles 2014) and usually involves the sharing of household space in close proximity and often, intimacy (Cudworth 2021). Despite this, the kinds of relationship between people and 'their' dogs varies significantly. The analysis of ethnographic material on which the chapters of this book draw indicate closeness between people and their dogs, yet the numbers of dogs routinely abandoned to shelters or the street are huge, with an estimated 664,000 dogs entering UK animal shelters each year, many of which will be euthanised (Cosgrove 2023). In addition, dogs deemed to have 'behavioural problems' are routinely killed by veterinarians at the request of their owners (Boyd et al. 2018). This is deeply concerning, given that dog behaviour which is often read as aggressive or vicious is increasingly understood as an expression of anxiety and fear. In addition, such behaviour and the often-lethal consequences of it, are driven by human expectations of how dogs should behave in a world designed for humans (see Donaldson 2005; Gurden 2021). Being a dog means being vulnerable to the whims of human owners and being caught up in dysfunctional human lives. Certain breeds of dogs are also additionally subject to risk due to racist and classist discrimination against dogs of a particular 'type' and the humans who own them (Guenther 2020; Kim 2015; Rosenburg 2011; Weaver 2021). While dogs certainly 'have their own lives and their own point of view' (Bekoff 2002, 24), and are agential and emotional, the agency they are able to express is highly limited.

Human households also exemplify different kinds of dog–human relations. Gary Varner (2002) suggests there are three models of incorporation: households having a relatively high level of inclusion, where human participants understand dogs as 'domestic partners'; households with restrictions on the use of space by dogs within the home where dogs are 'companion animals'; and 'mere pets' who are excluded from homespace (for example, chained in yards or caged). The people and dogs that populate the pages of this book have households organised in ways where dogs can be understood, in Varner's terminology, to be either companion animals or domestic partners. This said, even in more inclusive homes, human power is ultimately writ large into the macro (dogs being property in law, for example) and the micro (humans deciding the when, where and what of dog walking, socialising and eating, for example) politics of dog life.

Ultimately, this book will understand the lives of the subjects in this study to be posthumanist and captures this with the notion of 'muddied living'. In the processes of everyday routines and practices, this book will suggest that the human-animal boundary is muddied, compromised. Posthuman lives are understood to be muddy, but to varied degrees, as the maintenance of species boundaries is fluid and partial. On a more mundane level—humancentric notions of walking, making home and so on are muddied by being changed materially. Dogs are radically disruptive in human time and space by bringing in matter (mud, water and worse!) and asserting their will to bend places, practices and human desires and intensions to better reflect their 'point of view'. My use of posthumanism is associated with a concern with accounting for the more-than-human constitution of the social and the co-constitutive character of human/nonhuman lives. However, I also locate the book as *critically* posthumanist in its concern with relations in contexts of power (for a discussion of critical posthumanism see Cudworth and Hobden 2018). This introduction, and this book, will argue that intimations of a posthuman *condition* are very much overdrawn when human relations with other creatures are in the frame. We live in a world of species hierarchy, one in which human understandings and interests predominate. Dogs act, but as animal companions; they do so rarely in circumstances of their own choosing.

The final section of this introduction will contain the usual outline of the structure of the rest of this book and a note on some of the terminology used. I want to spend most of the time preceding this by thinking about what it means to be a dog, and to be a dog owner[1] in the context of human domination. Key to this discussion is an early contribution to animal studies in the social sciences, *Dominance and Affection: The Making of Pets*, by the geographer Yi-Fu Tuan (1984) with which we will begin.

DOMINANCE, AFFECTION AND THE 'PET'

Affection, Tuan suggests, is often apparent in social relationships, but

> affection is not the opposite of dominance; rather it is dominance's anodyne—it is dominance with a human face. Dominance may be cruel and exploitative, with no hint of affection in it. What it produces is the victim. On the other hand, dominance may be combined with affection, and what it produces is the pet. (1984, 1–2)

The pet, for Tuan, is the use of someone, or something, not for utility, but for the 'ends of pleasure, adornment and prestige' (1984, 2). Beings and things are rendered pets by being favourably treated, indulged and used as playthings. Power is endemic in the process of making pets. Pets can be humans, nonhuman animals or inanimate objects, but what is crucial is inequality, and this is sometimes obscured because affection 'mitigates domination, making it softer, more acceptable' (1984, 5). Whether a garden of topiary or a potted plant, a child, woman, slave or a 'fool', beings and things, through control and intervention by powerful humans, become pets. When it comes to animals kept as companions, Tuan highlights their disposability and replaceability, the ambiguity of species (with dogs being regarded as sources of meat in some cultures); the violent control of neutering and breeding wherein animals such as fishes have been manipulated into colourful ornaments for the home (1984, 88–99). Manageability and control, he argues, have been the key drivers for genetic manipulation; while in other cases, juvenile appearance has also been a consideration (in the case of Pekinese dogs, 1984, 105–6). While Tuan admits that humans may show affection and care towards animals, this is organised around human interest. I am often drawn back to *Dominance and Affection* precisely because it highlights the violence involved in making living beings and things into pets. Yet there is an arrogance in the certainly of the text and the presentation of examples which admits no ambiguity. There is also surprisingly little examination of the difference of kind of the creatures and things which are pettified, and their sociohistorical context. Perhaps I keep coming back to this text because I want more from it—to see how, precisely, the dominance and affection nexus operates.

Some insight into this can be found in the myriad of books on dog training. Patricia McConnell writes that 'Who dogs are and how they behave are partly defined by who we humans are and how we ourselves behave. Domestic dogs by definition share their lives with another species: us' (2002, xvii). McConnell considers that dogs' success as a species is a consequence of their domestication and their close relations with humans. Key to her warm and insightful books on dog training is the idea that 'miscommunication' between people and dogs emerges from a lack of understanding of species difference and species-specific behavioural cues—that dogs are canids and we are primates, and that this matters. Others have described this as a 'culture clash' which operates profoundly to the disadvantage of dogs (Donaldson 2005). Many training books incite their human owners to be more dog and to think more dog. For example, in *Unleashing Your Dog*, Marc Bekoff and Jessica Pierce focus on the sensory world of dogs, organising their book in terms of a discussion of dog hearing, sight, touch, taste and smell. With respect to the latter, they want owners to understand that

> A dog's world is a continuous cacophony and symphony of odors . . . dogs live in, and are consumed by, a world of smells. Our human sensory experience is dominated by sight, so to understand the world from your dog's perspective we really need to use our imagination. (Bekoff and Pierce 2019, 41)

The aim of the dog-training literature is to improve human-dog relationships and make them work. Yet as Haraway suggests, being a dog is a difficult job, requiring significant emotional and behavioural control (2003, 38). Bekoff and Pierce (2019) are more critical of the dog–human relationship than many. They consider that it is important to acknowledge that dogs are not naturally adapted to live with humans. Rather, in becoming a companion for humans, dogs are required to surrender many of the freedoms and pleasures of being a dog, and as Haraway suggests, to significantly curb their instincts. However much dog training is advocated as positive, reward or play-based, human control in the relationships remains key. The plethora of dog training material and estimates of the high numbers (72 to 85 per cent) of dogs with 'behavioural problems' (Powell et al. 2021) suggest a significant problem with our relationships with dog companions.

The enormous range of texts on dogs, dog behaviour and training all refer—often in a dedicated chapter—to what Mariam Motamedi-Fraser (2024, 2) calls 'dogs species story'. This story has its origins in the animal and behavioural sciences—in biology, ethology and psychology—and posits that while dogs are descents of wolves, the development of *Canis lupus familiaris* is a story of intraspecies relations, of dogs and humans, wherein dogs 'belong' with humans and would not *be* dogs without us. This matters greatly, suggests Motamedi-Fraser, as

> the story of how, precisely, a dog is understood to have become a dog contributes substantially to defining what a dog is expected to feel today; how a dog is expected to behave today; what treatment of dogs is justified today; and, above all, what about a dog passes almost unnoticed. (2024, 2)

Dog ancestry has been and remains contested and debated, the object of numerous studies, reviews and debates. A common thread, however, has been that dogs are descendants of grey wolves, and while the wolf has been the object of cultural demonisation and wolf populations victims of extreme violence and exterminist projects,

> the history of the dog appears as a linear success story. In a few thousand years, they spread across all continents following humans. They became the object of a wide variety of uses; a source of food and fur, useful working tool as hunter, carrier, hauler, guard, fighter, helper for disabled people, and loving companion. Dogs have become one of the most ubiquitous domestic species and the most common carnivore. (Lescureux and Linnell 2014, 233)

Tracing a genealogy of dogs' species story through evolutionary biology and ethology, Motamedi-Fraser (2024) argues that the story of dogs in the language of science is a story of belonging to humans, of 'following humans' as Lescureux and Linnell surmise. Despite differences in focus and perspective, the scientific story of the dog is that they are the original domesticates and their very specification as a species is connected to domestication, to household association—*familiaris* (see Motamedi-Fraser 2024, 62–90). Such a story of evolutionary becoming has focused on the concept of a natural bond between dog and human. Dogs, in being taxonomically defined by their relationship with humans, are understood as genetically hard-wired to 'be' with us. This, for Motamedi-Fraser, becomes 'the story of all dogs', and individual dogs bear the consequences of this. For example, she describes so-called problem behaviours in dogs as 'objections to their living conditions', which arise from the naturalising of the dog–human bond (2024, 5). As she argues, this story of 'the bond' places unreasonable and potentially highly oppressive burdens on dogs, who are subjected to ongoing and enforced socialisation in order that the ideal of the bond be both realised and attained (2024, 36–38). Thus, although the training of dogs has undergone significant changes in recent years, with positive behaviour reinforcement strategies aimed at 'realising the potential' of a dog (Charles et al. 2021), its linchpin remains obedience.

The dog is 'the most storied of all pet animals' (Fudge 2008, 10) and the stories science tells of dogs is, as Motamedi-Fraser meticulously shows, relentlessly humancentric. Erica Fudge goes as far as to ask whether a pet can actually *be* an animal, such is its' ambiguous status in being bound up with human ways of life (2002, 27). Others resist this humancentric story of the being of dogs, arguing for dogcentric understandings of what it means to be a dog. In attempting what they call an experiment in speculative biology, Jessica Pierce and Marc Bekoff imagine what they call a 'posthuman world', in a literal sense as a world after human beings are no more existent (2021, 3), rather than the sense in which it is used in critical social science (and in this book). They argue that dogs as a species are likely to survive and thrive without humans and question the notion therefore that the purpose of being a dog is to be a companion for humans; rather, dogs should be able to express their 'dogness' (2021, 13–15). While Pierce and Bekoff question the species story of dogs, Motamedi-Fraser wants to do away with it entirely. She sees it as engendered through racialised and classed frames, and with this I fully concur. She suggests that whereas race, conceptually, has moved from biological essentialism rooted in science and become understood as a social construct which is political, species remains biologically significant despite being a politicised social construction (see 2024, 205–7). Ultimately, her objection is that thinking with the concept of species erases the individual animal and fixes all dogs within a humancentric story of 'the bond'.

I consider, however, that we can retain species as a conceptual category without accepting dogs' species *story* and the political use of species and type. Rather, biological referents of species can be useful in specifying inappropriate uses and abuses of dogs (and other creatures) by humans. This also enables us to think about particular species' needs and suggesting what else might be more appropriate for other creatures and in terms of human relations with them. I also think that following Darwin's more expansive notion of species (which understands species differences in terms of degree, rather than kind, see Bekoff 2007, xviii) does enable uncertainty and destabilisation, not only in terms of essentialised categories but in terms of the permeability of the human–animal divide. In other words, while species is certainly a politicised social construction, it is also different to the biologically essentialised categories that have been used historically and continue to be used to suggest different categories of human beings. I am not convinced that our focus might be on the individual animal, but we can speak of collective interests of a species, or a variety of them, without fixing a species with a humancentric story. The challenge is to develop other stories, and hopefully ones that nonhuman creatures have a role in telling. This said, the politics of species has a problematic history and present that needs challenging. Here, what we require is an understanding of species as a socially constituted architecture of power which carries forms of social organisation—institutions and practices. We also need an understanding which both attends to human power and the ways this is intersected with intrahuman forms of exclusion and domination and enables contestation and other ways of being. In sum, we can retain species while questioning the social formations attached to its conceptualisation.

Tuan's analysis of the pet is ultimately a story of human power, even affection is simply a form of 'soft-power'. Is there no possibility to be otherwise in human relations with dogs? In feminist dog writing and wider work in feminist animal studies, conceptions of entanglement, empathy and co-constituted worlds are strongly present, and a relational conception of species that tells a very different story has been articulated. Lori Gruen (2015) for example, has argued for the further development of a relational notion of the self, incorporating responsiveness and responsibility for the well-being of all creatures, informed by 'entangled empathy'. For Josephine Donovan, understanding the 'qualitative heterogeneity of life forms' implies dialogical reasoning and the articulation of a nonhuman standpoint by feminist animal advocates (2006, 306–7). Feminist critique of animal rights theory focused on the failure to attend to the ways in which intrahuman forms of oppressions and exploitation, particularly those based on gender, sexuality and race, intersected with those in which nonhuman animals were caught. There was also a deep scepticism of the deployment of enlightenment rationalism in the development of a universal ethics for the human treatment of nonhuman animals by animal rights theorists (such as Tom Regan, see Rowlands 2009). Rather, for Donovan, it is via our attentive observation and our compassion, even for creatures who might appear alien to us, that we might enter 'dialogue' with nonhuman animals both as individuals and as collectives (as species). Donovan (2006) makes clear that attention and empathy must be accompanied by political engagement—through an analysis rooted in intersectional understandings of power and through advocacy which resists the objectification of animals and asserts their likely 'point of view'. In this kind of relational project, more respectful engagements and practices are envisaged not so much as an endgame, but as an emerging becoming-in-relation. As Fudge notes, 'conversation is at the heart of the human-pet' bond (2008, 2); and this is key to asking, 'How can we come to know this other being that is simultaneously in our home and utterly alien'? (2008, 1).

For Tuan, (animal) pets are the product of the development of sentimentality which accompanies nineteenth-century social transitions accompanying urbanisation (1984, 112). This social story of affection for pets has been one supported by sociologists (Franklin 1999, 85; Tester 1992) and the anthropomorphic presumption of pets as replacements for relations with humans seems to increasingly gain in traction in popular opinion (Harris Poll 2015) and in arguments that dogs are 'just like family' (Laurent-Simpson, 2021). I find this sociological pet story highly problematic, and Fudge nails the problem in considering that

> this is very negative. The idea that a pet is a replacement for a human, that living with a pet is as if living with another person, has within it the possibility of presenting the human-pet relation as somehow false, as a pretence, and that is not how many pet owners experience their relationship with the animals they share their homes with. (2008, 20)

Indeed, it is not. Yi-Fu Tuan, interestingly, reviewed *Adam's Task* by dog trainer and writer Vicki Hearne for the *New York Times* Book Review. Tuan (1986, 10) wrote that Hearne went 'out of her way to test her reader's credulity' in assigning human traits to dogs and horses. The review is a little vicious. It suggests that Hearne is 'excessive' in her anthropomorphism and also insufficiently self-critical in failing to see training as an exercise in authoritarianism. This is the hardness in Tuan's formulation in *Dominance and Affection*, for affection may as well be affectation, it is simply domination of a paler hue—with a 'human face', whatever that might mean. My reading of Tuan's view of Hearne is that it was written from a place of firm conviction of the superiority of humans. To see personality and emotional expression in animals, itself, is 'excessive' for him, perhaps.

'In 1992', writes McConnell, 'I fell in love with a dog named Luke' (2007, 236). Haraway (2017, n.p.), herself an admirer of the work of Hearne, describes Ms. Cayenne Pepper as 'a dog I fell in love with and who changed my life'. The stories shared by my research participants tell tales of friendship and companionate exchange, and of love and heartbreak. These stories of relations forged in everyday living together with dogs I take seriously and understand to be real, however much they may also be produced through dogs' species story of the bond with humans and relational structures of human power. What happens in human homes shared with dogs reproduces relations of species domination, but to various degrees and extents. Like Haraway (2008), I am interested in what else is going on. The stories people tell in this book reflect lives on the ground for dogs and their owners in which there are heterotopic moments, suggesting possibilities other than exploitation and domination. I have been drawn to Michel Foucault's (1986) somewhat underdeveloped notion of heterotopia, because he understands heterotopic spaces as both sites of contradiction and alternative possibilities *in addition* to reproducing dominant power relations. I have long been interested in how these alternate possibilities operate within social relations in which human interests predominate and nonhuman animal are dominated. In thinking through the social formations of species relations as those of domination, I developed the concept of anthroparchy. Ultimately, in the conclusion to this book, I will suggest that the spaces of muddied living in the lives of people and companion dogs can be understood to be heterotopic spaces in a nexus of human domination, spaces of contradiction and possibility in a relational system of anthroparchy.

ANTHROPARCHY—HUMAN DOMINATION

I started using anthroparchy as a graduate student, when looking for a concept to capture the social ordering of human relations with nonhuman nature. Anthroparchy literally means 'human domination'. In previous work on ecofeminist theory and in *Social Lives with Other Animals*, I argued that humans are entangled in complex systemic relations of domination with non-human animals (Cudworth 2005, 2011). Anthroparchy describes a system of social relations through which domesticate nonhuman animals are dominated through forms of social organisation (Cudworth 2011, 141). I conceive it is as a relational social system, a complex and relatively stable set of relationships in which the nonhuman living world is dominated through formations of social organisation which privilege the human and subordinate the nonhuman living world (Cudworth 2005, 63–71). I consider anthroparchy to have certain advantages over other possible terms used in the environmental and animal literature such as 'anthropocentrism' and 'speciesism'. The term anthropocentrism has been used to describe societies which are organised around a principle of humancentrism. However, 'centrism' underplays the severity of violence and exploitation involved in human relations with nonhuman animals and other beings and things, and so I tried to capture this with a term implying domination.

'Speciesism' has long been used by those concerned with the exploitative treatment of nonhuman animals (Singer 1990) and continues to have considerable traction in critical scholarly work on animals in the humanities and social sciences, and in political campaigning. However, it suggests a practice, a kind of behaviour, and is a parallel term to those describing other undesirable practices, such as ableism, classism, heterosexism, racism, sexism and so on. We do not (just) live in societies which discriminate against nonhuman species. Rather, we live in societies which are *systematically organised* around species hierarchy. 'Speciesism' carries various other problems with it—it is humancentric, it homogenises the vast differences across incredibly varied kinds of 'animal' within the concept of 'species' and does not attempt to account for different forms and degrees of exploitative relations that humans have with the incredible variety of nonhuman animals.

My conception of anthroparchy as a relational social system involves different forms and practices of power: oppression, exploitation and marginalisation, and I have used these terms to indicate distinct degrees (extents) and levels (amounts) at which social domination operates in the different formations it assumes. Some nonhuman animals can be said to be *oppressed*, for example nonhuman primates, mice and rats used for research in laboratories or incarcerated animals exhibited for human entertainment in zoos. Other species may not be implicated in anthroparchal relations, but exist in symbiosis, such as the biota in the human gut. Haraway (2003) is certainly right to suggest that what we think of as individuals are, at the same time, multispecies crowds. Different oppressive forms apply to different species due to their specific characteristics and normative behaviours, such as the presence of sociality and the ways in which this presents itself. *Exploitation* refers to the use of some being, space or entity as a resource for human ends, and one might speak of the exploitation of the properties of soils, or the labour power and reproductive capacities of domesticated animals, for example, while *marginalisation* refers to human centrism. In societies structured around relations of human domination, the complex and highly diversified nonhuman animate lifeworld is homogenised as nature, as Other to the human. Nature, as applied to nonhuman animals, is a socially constituted category with the physical referent of species difference.

Human relations with other species are constituted by and through social institutions and processes, and these can be seen as sets of relations of power and domination. I have suggested five sets of relations which are sub-systems of an anthroparchal social order. First, production relations, those emergent as humans interact with nonhuman nature as resources which are highly exploitative in the context of advanced industrial production and neoliberal capitalism. Second is domestication. While the breeding of plants and animals for human use is ancient, the last two centuries have seen intensification, such as reproductive interventions in farming, and the mass production of animals for the pet industry, for example, in puppy farming. Exploitation and oppression are endemic in mass breeding and overproduction and mean animals are disposable commodities. The third arena is political. Institutions and practices of governance may reproduce or alter relations of systemic domination, for example, through subsidising the farming of animals or making certain practices unlawful (such as, in the UK, hunting foxes with hounds, or bodily modification of animal companions, for example, shortening the length of a dog's tail). Fourth, there is systemic violence. This can operate similarly to violence affecting humans. For example, farmed animals may be terrorised, beaten, raped and killed. There are also slower forms of violence, such as habitat destruction which may push species to extinction and dangerous proximity to human settlements. Finally, cultures of exclusive humanism construct and police notions of animality and humanity, and encourage practices such as the consumption of animal bodies as food (see Cudworth 2011).

The term anthroparchy should not be taken to imply that all humans are in a position to dominate the nonhuman lifeworld, or that all humans engage in exploitative and oppressive practices all of the time. I have argued that species is constituted by and through, in collaboration with, human hierarchies—ideas of animality and of nature are vitally entangled in the constitution of race, gender, class and other human differences. We need to acknowledge the way species shapes the human and nonhuman lifeworld as both part of the *condition of life* on this planet and as a *relational system of power* (*Cudworth 2011*, 174–75). Intersection with other systems of social domination, of colonialism, patriarchy and capitalism, for example, means that some groups of humans are positioned in more potentially exploitative relations than others. In addition, individuals and collectivities choose not to exercise potential powers of domination and exclusion. While anthroparchy is a system of social relations therefore, it is also constituted by other relational systems and is the subject of contestation.

Domination takes different forms, shaped by social relations that are historically emergent with particular species, and species domination is co-constituted with other systemic power relations. In respect of dog companions, structures of oppressive power, and the keeping of dogs is bound up with the violent practices of the pet breeding, food and aspects of veterinary industries (such as the euthanasia of dogs for reasons other than alleviation of suffering due to illness or injury, or laboratory testing for veterinary pharmaceuticals). Dogs are nearly always part of anthroparchal structures of reproduction—the product of the breeding industries, and subject to its nexus of violence. As commodities, dogs can be passed from one household to another, or abandoned, if they are lucky, to shelters. Dogs are also caught up with forms of human violence both in the home (domestic violence matters) and outside (such as gang-related violence), which are gendered. The household is a site in which anthroparchal relations intersect with other systemic relations, of patriarchy, capitalism and other relational systems of social domination. Within the home, dog companions are commodities, still, in the United Kingdom, legally defined as property. The Western household is a key site of consumption, where the bodies of domesticated animals not considered pets may be variously stored, prepared and consumed as food by both human occupants and companion cats, dogs and others (Cudworth 2016b, as discussed in chapter 4). For many nonhuman creatures, a human home is a site of domination. Animals may be neglected or treated with cruelty or violence in this privatised space, relatively immune from public view.

Notwithstanding this context of human privilege, with its restrictions, exclusions and exploitative and oppressive practices, some animals kept as companions are able to bring their own selves into relationships with humans (Irvine 2004) and genuine possibilities exist for companionate exchange. As Haraway (2003) has argued, other relations and practices are found; and this is the primary focus of this book. In a human-dominant public world, the relatively privatised site of home, for example, might also be a site of resistance where the eating of animals who are not human or dog might be rejected, where relations of care and conviviality might enable the relative flourishing of humans and animal companions, and where practices are negotiated and human boundaries challenged. Human-dog relations are negotiated and developed through facial expressions, vocalisation and informative gestures and dogs read our behaviour (Masson 1998, 34). The long dog–human 'history of inter-species communication' (Alger and Alger 1999, 202) means that while we cannot know, we can certainly imagine what a dog might be feeling or experiencing. This is key to the development of caring perception and caring practices that later chapters of this book suggest are characteristic in households where dogs are included as either domestic partners or companions.

This book charts the practices of responsible dog owners who walk dogs regularly and have close bonds with them; who allow dogs into homespace and share that space enabling different degrees of freedom for dogs and different levels of physical intimacy. Despite this, the dark side of human relations with dogs emerges from the data, in tales of the past lives of rescued dogs, or the witnessing of human practices that range from carelessness and thoughtlessness to abuse, that were concerning for my human participants. It seems fitting therefore to provide some space in this discussion of human domination, for the consideration of systemic violence and its impact on companion dogs, and the ways this is encountered by some of the participants in my research.

DOG COMPANIONS AND HUMAN VIOLENCE

Even dogs, which are relatively privileged animals, are subjected to a range of violent practices. At the most serious level, dog companions are vulnerable to being killed. This is a vulnerability animals experience relative to humans, and some far more so than others, depending on their circumstances. For animals, different moral status enables their killing to be routine rather than exceptional, and to be normalised. Cary Wolfe has suggested that the human/animal distinction is continually reinforced because animals are sacrificeable, whereas humans, most usually, are not (2013, 9). He draws on Derrida's notion that 'beasts' (that is, nonhuman animals and animalised humans) fall below the law (Derrida 2009, 20–21) and thus might be subjected to a routine, large scale 'non-criminal putting to death' (Derrida 1991). There is a distinction in the language of killing that refers to differences in practice for domesticate animals—euthanasia for animal companions, as opposed to slaughter for animals which are killed to be eaten (Swabe 2009, 252). In both cases however, animals are seen to meet a non-criminal putting to death, and there are overlaps in the forms of killing. First, both are cold (objective and dispassionate) rather than hot (such as the 'impassioned' killing of hunting, Marvin 2006, 16). Anthropologist Garry Marvin observes that we have two models of cold killing: industrial and medical. In the medical mode, companion animals may be given a lethal dose of barbiturates by a veterinarian when their owner decides it is in the best interests of a suffering animal, or, and unfortunately often, when an animal is no longer wanted or judged capable of living the life of a pet. Certainly, medical killing differs significantly in motivation. There are veterinarians and owners of companion animals working hard to minimise pain, maximise quality of life and to secure as good a death as is possible (for a thought-provoking discussion, see Pierce 2012). Despite this, people can purchase dogs very easily and abandon them with few difficulties, discarding those no longer attractive or useful or who are ailing or aging.

The large-scale commercial breeding of companion animals relies on sophisticated marketing to boost demand, maximises profits through highly intensive and often cruel methods, and oversupplies animals for the pet trade to the extent that 'animal shelters are forced to destroy millions of unwanted pets each year' (Williams and de Mello 2007, 238–53). The treatment of some breeding animals mirrors the treatment of intensively farmed animals and is similarly dependent on gendered manipulation of fertility and sexuality and, on occasion, sexualised abuse (Dunnayer 1995, 14); while notions of breed purity and pedigree breeding narratives are constituted through strongly classed and racialised discourse (McHugh 2004, 91–95). Companion animals may be subject to domestic violence and may be mutilated, executed or otherwise harmed (Ascione, Weber and Wood 1997), while threats to harm or kill animal companions are found to be used as threats against women and children (Adams 1995). This impacts both human and animal survivors of abuse, both psychologically and practically (Flynn 2000; for a detailed discussion, see Taylor and Fraser, 2019). Estimates by animal welfare charities in Britain and the United States, indicate that numbers of companion animals mistreated (in organised fights, by drowning, stabbing, burning or by neglect) is statistically significant and seems to be on the increase in times of economic hardship (Wedderburn 2010).

The ethnographic work on which this book draws revealed high levels of anxiety amongst dog owners about a popular culture of machismo in which bull-breed dogs were implicated as weapons or trophies. Some people were concerned that their dogs might be attacked, and most were concerned about the breeding and treatment of such dogs. Tales of such dogs and human misanthropy are often interwoven in my data. The main field site for my ethnographic work was the Lee Valley Park, London's largest park which follows the River Lee (or Lea), running from the River Thames out through North and East London to Hertfordshire. When the East London sections of the river were dredged, hundreds of bodies of bull-breed, in particular Staffordshire terrier ('staff') and staff-cross puppies and dogs, were found.[2] One potential such case managed to find its way into my field notes:

> a white staff puppy, very small . . . they have it by 'mistake'. A client she works with came across some guys about to throw it in the river and drown it. It was the runt of the litter and they couldn't get the money they wanted for it.

Another case prompted much discussion amongst dog walkers when police were called because of a trail of blood found by an early morning dog walker. The dying dog had dragged itself under bushes having been stabbed with a knife more than twenty times. These two interviewees highlight the vulnerability of dogs subject to a non-criminal putting to death, but also, interestingly, use the word 'murder'[3] for the killing:

> It was a gang-dog, a kid's dog. That's what I heard. They thought there was a body in the bushes, and it was a body, but it was a dog body. Would the police have bothered if it was a dog? They came because they thought it was a kid. (F, London)

> Well, they're trophy-dogs and this is a trophy-kill. Well, I would say that poor bloody dog was murdered. It was murdered. They stabbed it to death for revenge. (F, London)

Media concern at the time on the weaponising of dogs prompted academic interest. For example, Jenny Maher and Harriet Pierpoint's (2011) research on young people and status dogs in South Wales found that of the young people they spoke with most owned dogs (more than half of which were bull breeds) and half belonged to a youth gang. Companionship and socialisation with friends were the main reasons young men identified for their dog ownership. Despite this, both they and youth practitioners also reported that dogs were kept for protection and enhancing the perception of their owners as tough or powerful individuals, and some admitted to using dogs as weapons and to organising dog fights. For Maher and Pierpoint, the abuse of dogs (and other small animals) and their neglect was paradoxical in terms of their value as companions. For my participants in East London, there were concerns about walking dogs and encountering people who keep animals for status or protection, and concerns about the use of dog walking open space for dog fighting:

> I've been involved in calling the police up here because there were gangs with great big sticks organising dog fights. You know, they're dog owners and they're organising dog fights. (F, London)

The breeding and treatment of bull breed cross dogs was of concern to my participants, and some linked this to a culture of machismo in which dogs were used weapons or "trophies" to enhance a street image:

> I work with a woman who's got a dog that she claims to be pit bull/Shar Pei cross—Colonel. And she keeps saying, 'He's a bit naughty,' and I go, 'What do you mean?'. She can't walk him by herself, he's too strong. He sees another dog and goes berserk. . . . You can't muzzle him because her bloke wants to walk him down the street and confront other pit bulls. . . . She's pregnant now, and she's petrified—what's she going to do with Colonel, who she can't control, and her baby? That's another one [dog] coming Battersea's way [referring to Battersea cat and dog rescue centre in South London]. (F, London)

> I think a lot of people buy dogs as trophies and they don't consider whether they [might] hurt another animal or a person and yet if they did [attack another dog] they're not covered, they're not insured, and they'll just, they'll probably just get rid of them [their dog], you know, and it's 'cause they don't [pause] they don't consider them something they love they just consider them something they can just replace. (F, London)

There was one occasion, however, where one of my participants queered the narrative of the bull breed dog as weaponised by young men. The owner of a huge animal, which he described as a mixture of Spanish mastiff, Rottweiler and English bull terrier, a "laid back" and "easy dog" who fit in well with a big and busy household, said that one reason for deciding to have a large dog was that he had

> three wimpy teenage boys at home, where we live's quite rough, er and I knew that having a big dog . . . would give them [his children] a bit of kind of local kudos and er some kind of protection . . . it's worked, er in spades really. The boys don't actually have to do anything, the dog just has to come out of our house attached to a big brass chain and that's it, you know, it's kind of nonverbal. Elmo makes a huge impression on the local idiots. (M, London)

More generally, the levels of systematic cruelty against dogs, and negative attitudes towards dogs, were of concern to most of my interviewees. Many had adopted dogs from rescue and rehoming centres and programmes of different kinds, and tales of various forms of human misanthropy are embedded in the stories of rescue dog lives from incarceration, starvation and routine beating to wounding with knives and cigarettes, for example:

> She was kept in a little cage in a back garden for the first four years of her life until they took her to the RSPCA. So, she doesn't like being left alone, she's quite needy in some ways. (F, London)

> She was half the weight when the woman first had her from the rescue centre—being half-starved has given her permanent kidney damage and a strange attitude around food. (Field notes)

> She had been returned to the [rescue] centre a number of times. One family only had her for the day. "How can you know a dog in a day?" he asks. When they got the dog, she had nightmares; she would wake up and run across the room to hide. If you raised your voice in the house, she would stand and shake and wet herself. (Field notes)

> He were in a state though weren't he [to partner]? He was a bag of bones. . . . He was locked up eighteen hours a day, with only a little window like that [demonstrates size with his hands]. That's all the light he had. When this bloke used to come at night-time, he'd let him in the garden, and he'd jump this seven-foot fence and be off down the road 'cause he'd been locked up eighteen hours a day. (M, London)

For some, other peoples' attitudes were deeply problematic. Several participants expressed the view that 'most people don't like dogs very much' and, in some cases, people they knew well considered dog companions to be disposable:

> think of your brother [to partner]. When we told him how much we would have to pay for her [their dog's treatment for her broken] leg, he said 'you should have had it put down and got another', and he's *got* dogs. (F, London)

A significant concern was the difficulties rehoming abandoned dogs of bull breeds with which animal welfare organisations have been periodically overwhelmed. UK animal welfare organisations such as the RSPCA (Royal Society for the Prevention of Cruelty to Animals) and Dogs Trust have campaigned to change the reputation of such dogs and for the amendment or repeal of the Dangerous Dogs Act (DDA) of 1991 which legitimates the mass killing of certain 'types' of dog. This legislation specifies unacceptable 'dog behaviour' and covers all breeds of dog but is also a form of breed-specific legislation. The act introduced a list of banned dogs—American pit bull terriers and three kinds of mixed mastiff breeds. Particularly poisonous is that dogs are assessed, by non-specialists, solely by appearance, as 'Whether your dog is a banned type depends on what it looks like, rather than its breed or name' (Gov.UK, banned dogs, 2023). If judged by a magistrate to have 'sufficient breed characteristics', dogs of these types must be destroyed (Defra

2009, 11), unless they are on the Index of Exempted Dogs and follow strict conditions such as wearing a muzzle in public places (Allcock and Campbell 2021). The legislation was hastily passed at the height of a media focus on 'dangerous dogs' which centred on attacks by American pit bull terriers. While amended by public order legislation in 2014, the breed-specific content remains and extension was made to the powers of authorities to seize dogs and on grounds of dog behaviour in private as well as public spaces. While the legislation has done nothing to reverse an apparent trend in attacks on humans by dogs, it has resulted in the seizure, incarceration and killing of large numbers of dogs, although statistics are difficult to find. In 2009 alone, 1,152 were seized by the London Metropolitan police (Barkham and Murphy 2012). At least 5,333 dogs have been seized by police since 2019, according to data provided by twenty-nine forces (but excluding the largest, the Metropolitan), and a total of 1,525 'dangerous' dogs were destroyed in the same period, figures released under the Freedom of Information Act show (Mercer 2021).

Police have simply to 'believe' that a dog may be of a banned type to seize and incarcerate (Defra 2009). Since the act became law, animal welfare groups have observed a rise in the breeding of American pit bull mixed breed dogs. These dogs are now often rejected by rehoming centres or taken in and destroyed. In many cases, their killing is compelled despite behavioural assessment by animal welfare organisations:

> Some dogs brought into our centres, as part of cruelty investigations, are later identified by the police as a prohibited type. Despite many of these dogs being friendly, well socialised, and perfect candidates for rehoming to responsible owners, the law doesn't allow them to be rehomed. (RSPCA 2014)

'Specific breed' legislation allows no consideration of individual cases or the context of human relations in which a dog finds itself. It has been called 'racist' in that physical features of appearance make individual dogs a target of the legislation (www.ddawatch.co.uk). The ways in which individual dogs are physically appraised certainly raises the question of the racialisation of species, as does the legal presumption that these dogs are prima facie legally 'dangerous' or 'vicious' by genetic disposition. Work in the United States has contested this stereotyping. Hearne (2007) was well-known for contesting the proposed killing of a 'vicious' dog, Bandit, who she eventually adopted, and the stereotyping of bull breed dogs has been linked to the racialisation of bull breed dogs and their owners (Kim 2015; Guenther 2021; Rosenberg 2011 and Montamedi-Fraser's discussion of 'black' and 'white' pit bulls, 2024, 207–14). Such critiques, tragically for the dogs and many of the people who live with them are concerned, seem to have limited traction. As I write this, the UK Government has added another bull breed type—the XL bully—to the list of banned dogs.[4] This has led to protests by owners and to objections from dog rescue and welfare charities, with concerns about the impact of potentially euthanising large numbers of healthy dogs on the well-being of veterinarians and staff working with and caring for dogs in rescue.[5]

There is a range of violences to which dogs are subject, some legitimate (killing by veterinarians and in the case of supposedly dangerous or banned dogs, by the police), others illegitimate (animals being killed, beaten or terrorised in the home) and often related to other forms of violence (domestic violence, gang-related violence). Whilst the killing of animals by lethal injection may indeed be in the interests of animals who are dying, terminally ill and/or in severe pain, large numbers are killed routinely in the breeding industry, by neglect and by veterinarians working with companion dogs and with dog rescue. The inability of owners, campaign groups and of animal rescue and welfare organisations to protect types of dog from culling under the DDA is illustrative of the vulnerability of dogs 'before the law'.

The behaviour of dogs, often related to their living conditions, makes them vulnerable. In 1990, my mother, who was a secondary (high) schoolteacher in North London, came home with a story of dogs running lose in the playground and inside the school building when she had arrived at work. The case was reported in national and local news and was a significant contributing incident to the hasty passage of the Dangerous Dogs Act.[6] The dogs involved in this case were Rottweilers, another breed which has been periodically demonised in the British press, most recently across much of 2022 (see, for example, Middleton 2022). Back in 1990, my mother was furious that the dogs were killed because of an attack by one, on a pupil, given revelations about the conditions in which the dogs had been kept (caged in a yard and not exercised). It is a pertinent illustration of the vulnerability of dogs that the punishment for behaviour by an animal in an abusive relationship should be death, but for their abuser the punishment of a mere fine and perhaps a ban on keeping dogs is considered sufficient. When first writing this introduction, I heard a dog walker suggest, in light of recent reports of dog attacks involving children, that Rottweilers be added to the list of dogs banned in the United Kingdom. The role of the media is crucial here, as Claire Molloy suggests, even small shifts in the discursive construction of dogs can be potentially lethal (Molloy 2011). As I finalise my text, it would appear that media coverage has removed the collective Rottweiler from the spotlight, and replaced this breed with the XL bully, a mixed(race?) dog with no breed standard and whose life or death will be established by a human judgement of a particular dog's appearance. This is lethal indeed.

The need for this discussion of the institutions, practices and cultures of violence in which dogs are caught, of their vulnerability to abuse by humans and of the pernicious nature of human power is because this is the backcloth for this book. The pet industry is an institutional system and set of production relations endemic to anthroparchal domination. These social relations are framed by law, culturally mediated and politically supported, for the most part. They are also contested through discourses of welfarism, rights and emancipation. Violence against animal companions is also co-constitutive of forms of complex inequality, bound up with relations of class, gender and race. While some extremes of violent practice may be de-legitimated, and some practices criminalised, the situatedness of dogs within the animal industrial complex of breeding, exchange, veterinary treatment and big pharma, and the socio-legal strictures of the use of public space frames the lives of dogs and those who live with them. This is in spite of how well people attempt to live with their dogs, and however far they are successful in enabling the dogs they live with to 'be more dog'.

WHAT'S COMING UP?

Recent empirical scholarship in sociological animal studies has begun to open to the presence of the other creatures that make up social worlds. Yet much of this research relies on standard practices and human-centred methods. In light of this, the first chapter, 'Research Pack', reflects on the challenges and contradictions involved in developing multispecies methods. The empirical research for this book is drawn from a project investigating everyday lives with canine companions via participant observation with dogs and their humans walking together in public spaces and fifty-two interviews with human 'owners' about their experiences of sharing their lives with dog companions. What was methodologically distinctive, was that most interviews involved dogs being present and most were mobile, or go-along interviews (Kusenbach 2003), undertaken while walking with dogs. An ethnographic diary was also kept in one field site, mapping events, interactions and routines of dogs, humans and others in the space of dog walking. This chapter focuses on how dogs and other nonhumans intervene in data collection and the challenges of reflecting dogs as lively agents in the data. It asks what is lost and gained by researching (literally) in the field and with a multispecies research pack. Haraway (2008) has used the concept of a 'pack' to describe humans and dogs living together, working together and walking together. This chapter elaborates the notion of a working research pack to suggest that if we want to be serious about bringing in nonhuman creatures into our projects, this means working with them and listening to what they might have to say. Taking animals seriously as subjects of research and as research actants, raises ethical issues, presents practical problems and requires imagination and resourcefulness.

Chapters 2 through 6 each consider the nexus of systemic domination and lively matter in the various practical aspects of sharing lives with dogs: walking, sharing home, eating, work and care and matters of relational identity in terms of emotional connections and kin ties. Chapter 2, 'Walking Edgelands', examines a routine, everyday and little-researched subject—people and dogs, walking together in public space. Dog walking is a timelised and spatialised practice where, in regular encounters, dogs and their human companions form relationships of community that might be understood as posthumanist. Such communities, the chapter suggests, are loosely knit, shifting and relatively tolerant of diversity and also distinctly located. The practice of walking through a particular space leads dog walkers to a knowledge of the places through which they walk and to the development of practices of care for those spaces and the creatures they encounter there—including other humans and other dogs. Research has already indicated that people who live with dog companions are more likely to interact with others in public spaces. Dogs have been seen to act as social lubricants and to encourage human participation in community (McNicholls and Collins 2000); but this underestimates the role of dogs as productive in the generation of relationships. This chapter suggests there are 'cross-pack' relations in terms of intraspecies and transspecies companionship and conviviality. Commencing with a discussion of concepts of community and posthumanism, the chapter locates posthuman communities in places that might be described as 'edgelands' (Shoard 2002).

Conceptions of home as a 'haven' have been critiqued on grounds of the elision of power relations (Rose 1993), yet home has also been understood as a place of resistance to, and refuge from, an exploitative and exclusionary public world (hooks 1992). Acknowledging differentiated relations of power and understanding homemaking as a process, chapter 3, 'Sharing Home', investigates the playing out of species' relations within home space. It argues that dogs transform domestic space through muddying human lives. This process is twofold. First, life in posthumanist households problematises boundaries between humans and other creatures in terms of relationships, behaviour and use of space. Second, muddied living involves breaching and maintaining domestic order. Muddied living is characterised by tension, power and compromise. Homes are posthumanist not just by including nonhuman animals, but through elements of dog agency in how home is made.

Pet food is an important element of the global political economy of meat and the animal-industrial complex (A-IC). In this context, chapter 4, 'Eating Animals', explores the role of carnism in the feeding of dog companions. Pet food is the subject of vigorous advertising encouraging human consumers to buy products that constitute a healthy, natural and/or economical diet for their companion, and thereby demonstrate care and responsibility. In such advertising, animal actors make their preferences clear and the consumption of this food—the bodies of domesticate animals who are not human or dog—is naturalised for animal companions. The analysis from my ethnography, however, suggests that many human participants did demonstrate concern beyond their animal, with the treatment of pet animals (dog fighting, animal breeding) and the (ab)use of animals in agriculture (particularly intensive farming). Although the majority of human participants and all of the dogs in the ethnography were meat eaters, many did place some restriction on what they ate in terms of the creatures they would eat, or the methods of production involved in the animal derived foods they ate. While many were troubled by the what we love/what we eat distinction, even the vegetarian dog owners and the lone vegan in the study all fed their dogs meat. Carnism (Joy 2010) was used unproblematically in discussing dogs' food and diet. Narratives of human choice infuse practices of dietary resistance to carnism, whereas biologism effectively reproduces the love/eat distinction when it comes to feeding animal companions.

Pets have been seen to provide security in an increasingly insecure Western world and compensate for human lack of affection (Franklin 1999; 2007), or be an outlet for human affection (Tuan 1986). Dogs have been described as surrogate children in their role as status symbols and emotional props (Redmalm 2014; Greenbaume 2004). Yet while dogs may provide affection, people resist the idea that pets may provide a substitute for relations with other humans (Charles 2014). In this context, chapter 5, 'Making Oddkin', considers what it means to be a dog companion and to be in a relationship with one. This chapter focuses on the stories people tell of their relationships with dogs in terms of emotional connections. While there are comparisons to be made with parenting, the stories from the human participants in this study showed that there was resistance to understanding the relationship with companion dogs as a form of parental guardianship, as well as aversion to the infantilising of dogs, who, rather, are seen as members of the household with a point of view. Different degrees of intimacy and affectional ties are present and varied levels of ambiguity and tension in the relationship people have with the dogs they live with. For some there is concern about dependency and lack of dog autonomy; for others, there is a special quality to the kinds of relationship we develop across the boundaries of species, which is odd, challenging and sometimes magical. Others resist the idea of *the dog*, stressing the individuality of their relationships with different dogs as unique individuals; while some consider that the dog is categorically significant, encouraging humans to think about themselves differently, and to 'be more dog'.

In discussing the relations between humans and dogs, Haraway (2003; 2008) considers the vulnerability of dogs in a situation where they are not understood to be ‘working’. She argues that respecting co-evolved working histories between humans and dogs may be a way to better ground our relations with dogs as a species. Yet there are ethical questions to be raised about working dogs and sociological questions about the meaning of work for both humans and dogs. While some of the dogs in this study went to work with their owners, none were working dogs as conventionally understood. Rather, they were mostly left, usually alone, at home while humans worked. This was a cause of anxiety for some people who made considerable accommodations to their daily practices in order that dogs were not left alone. This raises questions as to the role of the dog in the home that are often uncomfortable. The dog that waits alone for human(s) to return from work suggests that the purpose of a dog is to provide a welcome, some company or some other form of service. Haraway also suggests that the work of the dog is in living with humans, and certainly for some dogs this is an arduous task. The relationship between work and care is a powerful illustration of the intense ambivalence of the human-dog relationship, but is also an arena of hope, wherein both humans and dogs work towards better ways of getting along together. The chapter ultimately argues that key to a better relationship of exchange between dogs and humans is an understanding of carework as intrahuman, as undertaken for dogs and also by dogs.

The muddying of boundaries between humans and dogs raises questions about what it means to be a human and a dog in our times. As I suggested above, the spaces of muddied living in the lives of people and companion dogs are understood to be heterotopic spaces in a nexus of human domination. Within this frame, this book concludes by thinking through the implications of what can be glimpsed in these spaces of other possibilities, and what this might mean in terms of practical political interventions.

A NOTE ON TERMINOLOGY

It matters, as Haraway says,

> what we use to think other matters with; it matters what stories we tell to tell other stories with; it matters what knots knot knots, what thoughts think thoughts, what descriptions describe descriptions, what ties tie ties. It matters what stories make worlds, what worlds make stories. (2016b, 12)

So, before we proceed, a word on words and concepts. As seen in the discussion of anthroparchy, I do use species, and I understand it as a socially constituted concept of classification with meanings embedded in social hierarchy and with a biological referent. I also use the term animal in its biological sense. In Western cosmology, our animality is something from which we have increasingly sought to distance ourselves. As I have argued in my work with Steve Hobden, the uplift of humans from the animal condition is the key element of civilising projects and processes (Cudworth and Hobden 2014; 2018). It is worth noting that this is both a recent phenomenon and not a view shared by all of us as a species, as indigenous scholarship has long argued (Descola 2013; Vetlesen 2019). However, given the resilience of dominant imperial Western projects, Melanie Challenger (2020, 1) is right to suggest that 'the world is now dominated by an animal that doesn't think it's an animal. And the future is being imagined by an animal that doesn't want to be an animal'. This is a frightening prospect. To reflect the continued saliency and policing of the human-animal divide, and for clarity and ease of reading, I refer to humans and to animals but these terms should be read as meaning *human animals* and *nonhuman animals*.

After this introduction, I hope to largely abandon the word pet, except for when authors use it, or to retain the term for a concept, for example, for pet parents or pet food. As previously noted, I prefer the term owners for those humans who keep dogs in their homes as this reflects both the power dynamics and the unfortunate legal status of dogs. However, I think pet is a burdened concept—as Tuan suggests, it may refer to humans as well as nonhuman animals and even to plants and the landscape. For David Blouin (2013), to suggest that dogs are companions means that they are slightly different to 'mere' animals. Certainly, my research participants would probably choose *animal companion* as a descriptor over pet. I tend to use *animal companions* or *dog companions* to suggest companions who are animals or dogs, rather than the term companion dogs, as the latter suggests that the purpose of the dog is to be a human companion. In past work I have used Haraway's conception of companion species, however, I think this leads down trails where I don't necessarily wish to now go, and is in some ways not the best fit. In her earliest discussion of companion species, Haraway (2003) is interested in everyday, embodied and historicised forms of species relating and looks at the socio-historical cultural formations that emerge with dogs, specifically. However, the biopolitical constituent relations Haraway tracks with companion species do beg the question of whether any creature will do. Haraway has certainly shifted her position over the last twenty years (for an excellent discussion of Haraway on animals, see Timeto 2021). She is now less antagonistic to animal rights talk and sympathises with some critical animal studies scholarship (see Haraway 2016b, 230), and calls our attention to problematising human power. However, 'companion species' seems to underplay the unequal power dynamics therein. While the suggestion that neither humans nor dogs would be as they are today without the other might be generally the case, the evolution of dogs cannot be seen as a process of co-evolution in any sense implying mutuality. The history of dog breeding (discussed rather apolitically by the Haraway of twenty years ago) is a fine illustration of humancentric manipulation of the bodies and minds of dogs—resulting in the most diverse species of mammal on the planet.

My hope is that the stories told in this book, and the way they are told, will contribute in some way towards provoking a rethinking by human animals of how best to live with those we call companions, but of whom we are captors. There are those who live as best they can with dogs, and who imagine how life might be better for them and for those which they love—the 'dogs of their heart', to use Haraway's terminology. Some of the tensions and difficulties and some of the hopes expressed in what follows might prompt us to work on living better in our troubled present and to reach for ways of being entangled that enable dogness and promote flourishing of humans and dogs in shared lives.

Chapter 1

Research Pack

The boundaries of the social world have been stretched by scholarship in sociological animal studies, theoretically and empirically. However, much of this research, almost inevitably, relies on standard practices in both sociological research and in the representation of nonhuman beings and things. That is to say, nonhuman animals are included via humans talking and writing about them. We may include other animals in our projects, but we do so predominantly through standard humancentred methods. In this context, this first chapter considers the difficulties and possibilities of more-than-human methods, specifically the kind of ethnography on which this book draws. The ethnography provides the scaffolding for the empirical reflections later in the chapter and is drawn from my field notes of observations of the interactions between 'packs' of dog and human walkers and from interviews with those living with dogs. What was methodologically distinctive, was that the majority of interviews were 'go-along' or 'walk and talk' where, accompanied by two of the dogs with whom I shared my home, I walked with human interviewees and their dog(s), allowing for observations of human/dog and dog/dog interactions also to be recorded. Other interviews were undertaken at interviewees' homes or in public places where the interviewees' animal companions were present. In addition, there was an initial period of intensive immersion—an ethnographic diary kept for a calendar year in the London field site prior to interviewing. The diary recorded events, interactions and routines of dogs, humans and other beings and things in the space of dog walking.

At the time, when I was immersed in data collection, I considered that I was engaged in multispecies research, but the more I reflect on the data and how I obtained it, the more convinced I am that my practice remained humancentred. This chapter considers some of the ways in which conducting research with and about dogs alters both the data you obtain and the ways in which you understand the data collection process. This chapter moves through, first, a discussion of the animal turn in sociology and disciplinary resistance to it (Cudworth 2011a; Peggs 2012; Shapiro 2002; Wilkie 2015a) and, second, the implications of the animal turn in methodology (Charles 2016; Taylor 2012), in particular, in ethnography (Hamilton and Taylor 2017) before discussing the methods used in this project and some of the practices, events and issues that emerged.

ANIMALISING SOCIOLOGY, ANIMALISING METHOD

Some years ago, I was about to speak at a British Sociological Association (BSA) conference when an elderly white man stood up and enquired, very audibly, of those sitting in his vicinity, 'Is this the paper on animals?'. When it was confirmed that this was indeed the case, he declared 'Right, I'm off', and left the room not altogether quietly. For Rhoda Wilkie, this is perhaps an example of the worst kind of sociological reaction, exemplifying a 'rigidly minded' response to what is perceived to be an 'island of scholarship that lies outside of sociology's remit' (2015a, 326). Lest it be thought that such responses are the preserve of older, white, male British sociologists, another experience in a different panel at the same BSA conference saw an older, white, British, male sociologist full of enthusiasm for a paper on animals which I gave. In this case, however, the sociologist happened to be Ted Benton who made an important early intervention in sociological animal studies in the United Kingdom (Benton 1993). Since such pioneering work, and despite increased sociological interest, this, as Arnold Arluke puts it, 'can hardly be called a flood' (2003, 28). I still consider sociological animal studies to be often seen 'as mere boutique sociology . . . a passing fancy or trendy insignificance' (28–29). Thinking sociologically about nonhuman animals remains a peripheral concern.

Like other social science disciplines, sociology has generally understood there to be a fundamental 'discontinuity between humans and animals' (Noske 1997, 187). Considering species as a socially constituted category has proven difficult for a discipline whose boundaries were historically constituted around the designation of an arena for study—the social—which was defined as exclusively human (Cudworth 2011a, 20). Thinking about other animals has tended until very recently to be understood as at the fringes of, or tangential to, sociological concern (Whatmore 2006, 604). Most sociologists continue to write, for example, as if all household dwellers were human, taking little account of the empirical sociological work which exists on interactions between humans and the nonhuman animals who are so often to be found in the home (for example, Charles 2014 and 2016). Thus, it cannot be said—at least not yet—that sociological animal studies has moved 'from the margins to the centre' (Taylor and Twine 2014).

Moving 'to the centre' was a motif for feminist academic work for example in women's studies (Aaron and Walby 1991) and Black feminism (hooks 1984) and a comparison is invited. The gendering of the mainstream in sociology has been profound and there are grounds for optimism that the animal turn is here to stay. In examining the status of another interdisciplinary field, human-animal studies (HAS), Ken Shapiro locates it as 'solid, at the margin' (2002, 331). Yet since the early 2000s, we have seen the advance of interest in HAS evidenced in conferences organised, journals, books, papers and special issues in Australasia, Canada, Europe and the United States, such that the field has not merely expanded but 'exploded' (Shapiro and DeMello 2010, 307). It also would seem that youth is on the side of species inclusion. Leslie Irvine speculates that the future and success of animal studies in American sociology lies in the hands of 'graduate students who have the courage to challenge the field's outdated ideas about animals' (2012, 127).

Within animal studies itself, there are inevitably a range of perspectives and in particular, various kinds of politicisation (Jerolmack 2005); and animal-related issues can certainly engender politicised standpoints (Best 2010). As with feminist scholarship and women's studies, there is a continuum of positions and perspectives—from the more mainstream approaches of human-animal studies (HAS) to approaches linked to animal advocacy and different kinds of critical theory, critical animal studies (CAS). This has also led to further political divisions as can be seen with the generation of radical animal studies in reaction to what is perceived as mainstreaming in CAS (see Best 2010 and 2014; Nocella and Socha 2022; Wilkie 2015b). Elsewhere I have suggested that feminist approaches constitute both a distinctive and also a variegated field of positionalities that deserves consideration—feminist animal studies (Cudworth 2016a; Cudworth, McKie and Turgoose 2023). As Kay Peggs (2014) contends, the placement of different approaches on centre-margin continuum will be varied. Such apparent division and discrepancies in approach involve the mapping of new territories of empirical scholarship and analytic dispute and the literature is gathering apace. Limping behind somewhat, is the discussion not so much of what we study and how we think about it, but how we approach it.

Multispecies studies compel researchers to go beyond their disciplinary boundaries (Birke and Hockenhull 2012). The challenges of multispecies research have already led some to consider that HAS and multispecies work may be able to transform orthodox 'methodologies that hitherto were imbued with anthropocentrism' (Segerdahl 2012, 156–57). What methods might be open to capturing 'co-habitation, co-evolution, and embodied cross-species sociality' with a multitude of critters? (Haraway 2003, 4). In researching the lives of people and dog companions, observing people and dogs, talking to people about their lives shared with other creatures and their engagements with them seems an obvious starting point. Yet this kind of ethnography is necessarily problematic. We human subjects are those who decide what and how to collect data and how to present it, and this inevitably involves representation; a representation that is more difficult when those we represent are not human and often share a tiny amount or nothing of our language (Sanders and Arluke 1993). We observe, record, speak and write, and it is unsurprising therefore that most usually what we find begs 'the obvious question, where are the animals themselves in this research?' (Hamilton and Taylor 2017, 2). There are also ethical questions raised—even if we manage to represent animals in our research, what do those represented (either individually or collectively) actually gain from our enquiries and academic outputs? (see Gaard 2012).

The use of ethnographic observation and interviewing in field sites where humans and other creatures interact has used traditional sociological methods effectively in capturing the ways in which humans and other animals make relations and interact. Sociological ethnographic work has been undertaken in various contexts such as animal shelters (Alger and Alger 2003; Irvine 2004), veterinary clinics (Sanders 1993), living and playing with dog companions and dog training sessions (Arluke and Sanders 1996; Sanders 1999). Such studies have been important in demonstrating the ways in which animal practitioners and owners of animal companions understand the animals they work and/or live with as persons with their own views about the world. Janet and Steven Alger's work on human-feline relations emerging from their ethnography of a cat shelter used observation and interviewing show how cats interfere with human activities to make their preferences known, exercise choice (where to eat and sleep, for example) and display emotional states (Alger and Alger 2003b). Irvine found that would-be adopters of animals attempt to communicate with the animals they encounter and read relations into those encounters:

> The dog then turns over, stands on her hind legs, and places her paws on the woman's thigh. The woman brings her face so close that the dog can lick her. The woman laughs and talks softly to the dog for a minute, saying, 'Would you like to come home with me? Do you think we can be friends?' (Irvine 2004, 90)

From such observations, we gain an insight into the ways in which co-becoming of interspecies relations is developed. Irvine's combination of ethnographic observation (in both a shelter and in dog parks) and her interviews with those potential adopters of animal companions gives insight into the texture of lives people live with animal others. Irvine suggests that close attention we pay to animals enables us to interpret behaviour and physical cues and to communicate at some level. Through reading animal behaviour and practices of interspecies communication, we can imagine what it is like both to be a dog and how a particular dog might be feeling. It may be that what Irvine, the Alger's and I have researched is 'easy pickings' because the possibility of reading across species is more straightforward when it comes to domesticated mammals, particularly those with which we are very familiar, as with cats and dogs. Yet of course, we may inevitably have read everything wrongly for nonhuman animals also behave in ways that are sometimes counterintuitive to us, concealing, at least to human eyes, that which we—as humans—are often able to reveal to each other partially or wholly (Hamilton and Taylor 2017, 8).

Ultimately, it can be said that the policing of boundaries between human and nonhuman remains, and such research can be seen to 'simply maintain dualist conceptions whilst moving the boundary slightly (that is from human/social v. animal/natural to human and (some) animals/social v. natural' (Taylor 2011, 206). This is because the implications of such research suggest that animals that we humans know as individuals or in groups in social settings enable us to better appreciate the ways in which animals are more widely embedded in our social networks and practices. As Lindsay Hamilton and Nik Taylor note, multispecies ethnographies can sometimes not help but 'fall into the trap of prioritising human knowledge' (Hamilton and Taylor 2017, 8). Some working in animal studies in anthropology, such as Samantha Hurn (2010), prefer the term 'anthrozoology', to human-animal studies as she sees the role of anthropological enquiry on the interactions between humans and animals as an important manifestation of what it means to be human. Certainly, this foregrounds the role of ethnography in examining forms of human interaction with animal Others, and for Hurn (2012, 207), seeing minded animals as agential actors in research projects is both acceptable and desirable. Recent years have also witnessed an emerging literature on multispecies ethnographies (Kirksy and Helmreich 2010); that consider the involvement of animals alongside humans in a range of social practices (such as work, sport, farming, research, or companionship see Charles and Wolkowitz 2019; Haraway 2008; Bear, Wilkinson and Holloway 2017; Taylor and Fraser 2018). This has accompanied calls for methods better able to attend to human–animal relationships (van Dooren et al. 2016) and that attend to the 'voice' of animal through visual and sonic data (Birke 2014; Taylor and Fraser 2019). Veissier and Miele (2014) have suggested that genuinely multispecies research needs multi- or interdisciplinary approaches involving research teams with expertise in the science of animal behaviour as well as social sciences (see also Birke and Hockenhull 2012). I have much sympathy with this suggestion and am conscious of the human bias in my own project. Despite such caveats, in terms of my own data, I do think that ethnographic observation, conversations and interviews do enable nonhumans to be present, at least to some degree.

For those who identify with 'multispecies' scholarship rather than human-animal studies, this does not go far enough. While HAS has focused on people's relationships with animals or a particular species, just as my own ethnography has focused on people's relationships with dogs, 'a multispecies approach focuses on the multitudes of lively agents that bring one another into being through entangled relations' (van Dooren et al. 2016, 3). I am not convinced that differences in the field are so marked. The common thread in 'multi-species scholarship' is 'understanding what is at stake—ethically, politically, epistemologically—for different forms of life caught up in diverse relationships of knowing and living together' (van Dooren et al. 2016, 5). Yet I do not think that those working on relations between humans and a single other species would disagree. A key methodological innovation here is taken from Anna Tsing and her notion of 'passionate immersion' (Tsing 2011) which '[a]t its core it involves attentive interactions with diverse lifeways' (van Dooren et al. 2016, 5). The difference may really lie in the scope and that multispecies work is often looking at many, or more than two species in interaction. Anna Tsing's (2015) work on the matsutake mushroom draws in mushrooms, trees, humans and relations of migrant workers, corporations, forests and their management, for example. In my own ethnography, other animal companions are drawn into the frame in addition to dogs, cats, sometimes birds, guinea pigs and rabbits. Relations with other creatures within specific places, such as non-domesticate animals encountered in spaces of walking, are also a feature of the data. Another anthropologist, Tim Ingold, defined 'ecological anthropology' as a subdiscipline that contextualises how 'human beings and other organisms are bound in webs of life' (Ingold 2012, 428). Might our encounters in the more-than-human world be best brought into written form as ethnographies that give nonhuman others vitality, presence, perhaps 'thickness' on the page and in the minds and lives of readers, ask Thom van Dooren and Deborah Bird Rose (2016, 84–85)? Yet as Helen Kopnina (2017, 347) argues, this alone is insufficient, for particularly where the suffering of nonhuman creatures is concerned, observation and thickness of presence cannot do all the work for us—a standpoint is also required. She concludes that while 'the rhetoric of entanglements is sympathetic to nonhumans, it remains largely conventional' (2017, 351).

Despite such caveats, whether scholarship identifies as human-animal or multispecies, ethnographic observation and conversations or interview styles related to it do enable humans and other creatures to be present, even if we cannot, ultimately, get away from ourselves as human researchers. Ethnographies might invite curiosity about other ways of life (van Dooren and Bird Rose 2016, 86) and enable us to better understand its fragility and (often shared) vulnerabilities. They also pique our curiosity, for while I had walked with dogs for many years, it was only when I wrote about what I had observed that I realised that I had not really looked before. Marvin (2005, 4–6) might explain this in terms of the different degrees of attentiveness required by seeing, looking, watching and observing. While I had years of seeing dog and human walkers (registering their presence) and perhaps looks at them (taking more than a passing glance) I had not really 'taken notice' and paid the concentrated attention of the observer. In addition, ethnographies are not necessarily impartial. Generating empathy may chime with advocacy for nonhuman animals 'as part of the emancipatory and critical endeavour of posthumanist scholarship' (Hamilton and Taylor 2017, 174). Certainly, an emancipatory agenda was my ethical project in relation to my research. I intended not only to map the ways in which human and dog lives are entangled and altered in species interaction but also to consider the difficulties experienced by dogs and their humans as a result of the property status of dogs (including the documenting of histories of abuse in the lives of adopted dogs), restrictions on the movement of dogs in public space and the challenges to 'living well' as companion species.

WALKING WITH AND TALKING DOGS

The ethnography I undertook used two methods of data collection: semi-structured interviews and participant observation. The first phase of the research was a period of intensive immersion in one of the field sites where daily observations were undertaken and field notes were kept in the form of an ethnographic diary of encounters with dogs and their people in part of London's Lee Valley Park, for a calendar year.[1] This gave a detailed picture of the nature of this particular community of dog walkers and their interactions over an extended period and also enabled data to be collected on the impact of the seasons, weather and other creatures and forms of life—for example, the way in which snow, flooded marshland, long grass, cold or warm weather impacts the kinds of interactions between humans and dogs. The diary recorded events, interactions and routines of dogs, humans and other beings and things in the space of 'dog walking', noting interactions between individual dogs and humans and interactions within and across walking 'packs' (of human(s) and dog(s)). While my reflections were also recorded, the diary data was not intended to be used autoethnographically (albeit that scholarship on relations with dog companions has utilised this as method, for example, Satama and Huopalainen 2019). In addition, material is drawn from semi-structured interviews with dog owners, investigating their relationships and everyday lives with canine companions.

The interviews took place in two phases, initially in London, which was intended at the outset of the project, with a single field site in the Lee Valley

Park. A second phase of interviewing was undertaken in a rural field site, when this opportunity presented itself. The majority of interviews were 'walk and talk' mobile interviewing where I walked, along with my two canine research assistants, with human interviewees and their dog(s). This is a combination of ethnographic participant observation with interviewing which has also been called 'go-along' interviewing (Carpiano 2009). The idea is to accompany informants as they go about their daily routines, asking questions about their lives along the way (Hall, Lashua and Coffey 2006). Some using this method are very interested in the effect of space on the nature of what people say and thus tend to have fixed routes (for example, Jones et al. 2008). However, I followed whichever route my participants chose, wanting to be interviewing people about their relationships with their dogs while they were walking their dogs on their usual or preferred route. My priority was to put people at ease and to talk to them about their dogs in a situation where they are dog-focused. This interview approach allowed for observations of human/dog and dog/dog interactions also to be recorded and, in some cases, enabled dogs to have an influence over the route and the timing of the process. Some interviewees chose other locations, such as pubs, cafés or their homes, or relatively rarely, my own home. In all cases, dogs were present.

Thirty-seven interviews were undertaken in 2010 and 2011 with people walking dogs on the marshes of Walthamstow and Leyton (in the main) and also Hackney and Tottenham, which form part of the Lee Valley Park. Relocating to a village in rural Leicestershire meant I became part of another community of dog walkers and enabled me to test the extent to which locality affected the data. Thus, a second phase of interviewing was undertaken in 2014 where fifteen interviews were undertaken with people walking dogs in and around the village.[2] There were, however, no substantial differences in data obtained from interviewees living in urban and rural locations; an unexpected finding.

I was drawn to walking interviews as a way of both maximising authenticity and trying to include dogs in my project. While it has lately attracted attention as part of the turn to mobile methods, walking as a research device is not a recent innovation (Bates and Taylor 2017). In an important early paper, Margarethe Kusenbach (2003) argues that go-along interviewing has the potential to access some of the transcendent and reflexive aspects of lived experience. She draws on urban ethnographic fieldwork in arguing that mobile interviewing enables the teasing out of complex and subtle meanings of place in everyday experience and practices. Kusenbach (2003, 462) argues that observation alone cannot access peoples experience, while regular 'sit-down interviews' take interviewees out of the environment where the activities you want to research are taking place. In both study sites, I was researching my own community of dogs and humans[3] and thus the go-along interview, combined with the kind of intense observation of the space recommended by Kusenbach, seemed an excellent choice because it enabled research in a space where people usually encountered me. What I undertook were what Kusenbach recommends as 'natural' go-alongs which have the key advantage I was looking for, a greater degree of authenticity than might be possible in unfamiliar settings:

> For the purpose of authenticity, it is crucial to conduct . . . 'natural' go-alongs. By this I mean go-alongs that follow informants into their familiar environments and track outings they would go on anyway as closely as possible, for instance with respect to the particular day, the time of the day, and the routes of the regular trip. (Kusenbach 2003, 463)

As my sample of humans and dogs were necessarily those who walked with dogs and who undertook what they referred to as a 'proper walk' (that is, not a walk on streets and a walk that is relatively long), the time window—of between forty-five and ninety minutes—ensured no disruptions to the regular routines of human-dog walking packs and was also happily what Kusenbach refers to as the 'optimal' time for a go-along encounter (2003, 464). The only disruption to routine was for my research assistants and for my older dog Kevin (aged thirteen when the interview phase of the project began in London)—although fit and enthusiastic, I did wonder whether I was asking too much.

The two field sites suggested that there are specific spaces of dog walking—edgeland spaces—that are different to other venues where dogs and dog owners may encounter each other and hang out. In the United Kingdom, many urban-dwelling dogs are walked in public parks or around streets. Dog parks, or fenced areas of public parks where dogs may be off-leash (or lead) are relatively rare in the United Kingdom (compared to the United States, for example); although lead-only dog walking is increasingly specified. Edgeland spaces, however, are not subject to the levels of surveillance apparent in public parks and enable off-lead walking. In both study sites, paths across fields or meander paths (small pathways, made by walkers using a repeated route) around fields, and on marshland meant that dogs could walk relatively freely. Certainly, the human walkers in this study saw edgeland spaces as those of relative freedom for dogs, as one put it: 'It's not a proper walk if I don't bring them [her dogs] here [the marshes of the Lee Valley Park]. They love it here because they can do their thing, run about with their people [other dogs], rootle around'. The next chapter will consider *edgeland* space in more depth and argue that this enables particular kinds of relations between packs of humans and dogs to form.

The dog walkers of a village in rural Leicestershire navigate fields with farmed animals and crops, crossed by official routes and meander paths. Movement along paths leaves room for some interaction, but it is in places of congregation—by hedges at the sides of fields, on farm tracks or in less managed areas of recreation grounds—that people and dogs gather and hang out together (see figures 1.2 and 1.3). The Lee Valley Park, London's largest park, is a patchwork of more traditional marshland that has been upgraded in some sections with paths for walkers and cyclists. It is seen by some of its users as 'countryside' and there is an abundance of wildlife with parts of the park designated sites of scientific interest and provided protection as nature reserve. It is also very much an urban wetland—viaducts crossing the park are decorated with graffiti and parts are edged by housing estates (see figure 1.1). There is much space for meeting and stopping and the relatively open landscape encourages gatherings of packs who know one another. Kusenbach describes the 'go-along' as a form of 'deep hanging out' (2003, 265) which perfectly captures elements of my roaming interviews, although my hanging out was more directed—while I followed the routes of the packs I and my research team accompanied, I had my questions invisibly filed—waiting for the best moments to slip them into our conversations.

A wide variety of dogs were involved in the study including Airedale, Jack Russell, Staffordshire and Norfolk terriers, schnauzers, German shepherds, varieties of sighthound, spaniels, collies and retrievers, designer mutts (mixed breed) such as goldendoodles and cockapoos and multivariate 'proper mutts' such as one described by my interviewee as 'a typical British mutt circa 1972, a collie/lab sort of thing'. It was notable that in terms of the breed of dog, the London field site proved far more eclectic. The range of breeds was wider, the number of mixed breed dogs larger, and in particular there were large numbers of sighthounds.[4] In the rural field site there were large numbers of terriers, but far more dogs were pedigree breeds, and there were significant numbers of Labrador retrievers, spaniels and designer crossbreeds, particularly cockapoos.

The human participants had a wide range of occupations[5] and crossed divisions of class and wealth but were overwhelmingly white and mostly female. Similar to studies considering animal companions in the context of the 'family' (Charles and Aull Davies 2008) or within the home (Fox 2006), I found more women than men to interview, particularly in the rural study site where twelve of the fifteen interviews were with women, and three men were interviewed together with their female partners. In part, this is because the majority of dog walkers observed in both locations were female, and because more women than men were willing to participate, particularly in the rural study site. In chapter 6, it will be suggested that this gender disparity may be due to walking with dogs being a form of gendered carework. Readiness

Figure 1.1 Housing bordering the River Lee (also spelled Lea) and Walthamstow Marshes.

to be interviewed may also be because women are likely to be less reserved about speaking about their everyday lives and homes, or the qualities of their relationships, a finding of previous studies on the role of companion dogs in families (see Charles and Aull Davies 2008). The overwhelming majority of interviewees were white, a reflection of the dog walking population in both locations. The social class background of the human participants is difficult to ascertain (except where interviewees explicitly discussed their background and/or occupation, and/or where in the London context, class might be inferred from accent).

The observational material and the interviews chart the practices of responsible dog owners who walk dogs regularly and have close bonds with them. All participants in this study shared homespace, and Varner (2002) would describe these households as those where human participants understand dogs mainly as 'domestic partners' or 'companion animals' but not 'mere pets'. Some of the dog owners who participated in this study specifically referred to themselves as 'dog people'. As one put it, 'There are people who have dogs and then there are dog people' (F, Leicestershire). The distinction being made here was that some people do not necessarily have an affinity with dogs but may have a dog with a specific purpose in mind, for example, that it would

Figure 1.2 Field edge with meander path made by regular walkers in the rural study site.

be beneficial for children to have a dog in the home. 'Dog people' would be those who identify closely with dogs. In the case of the woman quoted above, they would be those who try to 'understand a dog on a dog's own terms' (F,

Figure 1.3 Research assistant Ruby on a farm track

Leicestershire). These were also people who as a result demonstrated tolerance of a variety of dog behaviour, who, for example, 'don't take *personal* offence if your dog barks at their dog' (M, London). There were different levels of interaction between groups of walking packs and different relationships emerge between different animals in their interactions, but whatever the strength of bonding, I did find that dog walking communities are always constituted by and through dogs as well as by their human companions, as we will see in the following chapter, and dogs, as previous research has found, did lubricate social encounters between humans. It is fair to say that the data for this project is also co-constituted as we were a research pack. This certainly changed the data I was able to collect and imbued it with specific features.

As will soon be apparent, human laughter is a feature of the interview data. In their study of the changing nature of family in which people proffered material on their relationship with animal companions, Nickie Charles and Charlotte Aull Davies (2008) considered the use of laughter by interviewees to be an attempt to lessen the impact of including animals in their discussions of family, or to reflect ambivalence about revealing too close a relationship with their animals to researchers. In this project, some of the laugher may be due to such embarrassment, but also it may simply be that, as one interviewee put it, dogs are 'just hilarious' and that retelling stories of lives lived with dog companions makes people laugh. At some points in the interviews, people had to pause because of laughing, or laughed so hard that they had tears in their eyes. Writing this chapter, and indeed this book, has often made me smile.

RESEARCH PACK

I do not think that it is an exaggeration to say that I worked in a research pack, accompanied by the dogs who share my home. The interview sample was actively shaped by the relationships of my dogs, to others (both human and dog). I began the research in London, working alongside two dogs. First was Gypsy, an easy-going female whippet-lurcher who had lived with me since a puppy and was generally accommodating of all dogs and humans. She was accompanied by Kevin the Jack Russell terrier, a small man who considered himself to be big. Kevin was effectively rescued, being passed over the fence by a neighbour who no longer wanted him and was threatening to take him to a rescue centre. He came with a food bowl, but no collar or lead. His nervous barking from the other side of the fence was one of the factors which prompted us to buy the house we had in East London—of the many we looked at, we remembered which one it was, the house with a dog named Kevin next door. Once we had moved in, he spent much of his time on our side of the ineffectual fence and seemed to like hanging out with Thumbelina the cat before he came to live with us. Kevin and Gypsy were my research assistants and came with me on every walking interview and to those conducted in other locations in the London phase of the study—usually people's homes or otherwise public spaces, specifically a pub and a café both located in the field site. On the occasions when people came to my home, the dogs were always present. The dogs' involvement was crucial—they secured legitimacy in the field and interactions between people and dogs stimulated and provoked responses in interviews (see also Lane 2015).

In the second phase of the project in rural Leicestershire, losing the aged Kevin had meant a new dog in my life, Ruby, to whom I often refer as 'the dog that was sent to try me'. While Gypsy was a relaxed soul, at ease in the world, Kevin and Ruby were/are of a nervous disposition and had/have very clear likes and dislikes—of both certain other dogs and certain other dogs' human companions. Ruby, in particular, is highly 'reactive' and a combination of great affection and significant nervous 'aggression' towards both some dogs and some humans. She tries me because, as a friend described, she is 'so resentful of human control'. She would certainly be a dog who Motamedi-Fraser (2023) would claim is 'objecting to her living conditions', for at least some of the time. Sadly, for Ruby, in a humancentred world, some level of obedience is necessary for life. Both the interview sample and what could be observed is shaped by the relationships of Kevin and Ruby to others. For them to be involved, I did not walk and talk with packs of dogs and companion humans whom my dogs disliked. This would have been stressful for all concerned and was the most significant factor in the selection of the interview sample.[6] Karen Lane has suggested that 'a dog perceived as an aggressive dog would make a poor research assistant for collecting stories' (Lane 2015, 35). However, unlike Lane's mobile ethnographic work with her

Figure 1.4 Research assistants Kevin and Gypsy on Walthamstow Marsh

friendly and affable wheaten terrier, my research assistants were not expected to engage with all humans and dogs as they encountered them but were involved in selecting the sample. Lane muses as to whether it is the breed, appearance or personality of the dog, and their relationship with the human that makes the research pack work but working with three different dogs with different personalities led to different kinds of engagements and data, which in my view enriched the findings.

Gypsy, Kevin and Ruby were not 'official' research dogs, as was the late Ms Cayenne Pepper who gained research status at the University of California Santa Cruz in order to be able to accompany Donna Haraway to her office (Haraway 2008, 206). Haraway argues, however, that while Cayenne's official status was about securing her company on campus, the real research role Cayenne assumed was in living with Haraway. In the work of training, dogs and humans learn, Haraway says, how to 'play with strangers' (2008, 243). This understanding of co-becoming through living together enables Haraway to assert that in producing the chapter on agility, play and training for *When Species Meet* it was 'not a literary conceit' to say that she wrote the chapter with Cayenne but, rather, 'a condition of work' (2008, 205). Perhaps living with a dog whom I know resents aspects of her living conditions I am very sensitive to the power relations of training in ways Haraway is not. In this project, Kevin, Gypsy and Ruby were field researchers, and their behaviour sometimes perhaps illustrated my own 'deficiencies' in training. In discussing dogs as ethnographic research assistants, Lane reflects on occasions where her dog breached dog walking etiquette and prevented better engagement with a group of people walking with their dogs (Lane 2015, 35). More commonly for Lane however, dogs can precipitate contact with people and act as lubricants to more effective encounters, even supporting increased trust by participants, in the researcher (33). Similarly to Lane, I prefer to understand the interventions of dogs in research sites as the point at which my canine assistants 'become field actors and condition an encounter' (31). For example, at times, my research assistants chose to interrupt an interview with a human by barking, so that transcripts are frequently peppered with:

> Interviewee: Shhh!
>
> Erika: Kevin!
>
> Interviewee: We're doing an interview here, Kevin.
>
> Erika: Kevin! Enough! Sorry about that, he barks for no reason sometimes, he seems to just enjoy it. Where were we?

In this case, Kevin's intervention conditions the encounter by changing the topic of discussion. This begs the question as to whether insights were lost or gained in the change of topic. In other cases, my research assistants appeared to have their own research agendas. On occasion the lovely Gypsy would disappear to sample some of the smorgasbord of delights (decomposing critters, various kinds of poop) offered in our field site so that interjections of 'come *on* Gypsy!' or 'hang on a minute she's eating crap again' were common. On occasion, being a relatively young dog (four years old) at the start of the project, she would decide on her own route with the dog we were accompanying:

> Erika: Where have they gone? [Pause.] There they are over there—got their own agenda there.
>
> Interviewee: Yeah, sneaking off together even—one minute he's pretending they don't like each other and now they're gonna elope.
>
> [Some time spent going back and calling, a slight change of route.]

Alternatively, she might set out on her own:

> Erika: Oh, there she is. Where have you been? She's frightened of that bridge as well. She doesn't like crossing it.
>
> Interviewee: Is she alright with the other one though? [Followed by a digression discussing the various bridges across the river and which dogs dislike which ones.]

The above encounter changes the route of the walk and leads to a discussion of forms of anxiety in dogs which the interviewee was attempting to grasp in her own relationship with her dog. When Haraway (1997, for example, and others) speak of dogs as co-producers of knowledge in the laboratory as subjects of scientific experiments, it unsettles me greatly. This is because the extraction of knowledge derives from the forced labour and suffering of animals rather than anything remotely like a joint endeavour. Here, however, wandering or sitting around with other humans and other dogs meant that the data is slightly dog-shaped. At the time, I sometimes considered dog incursions into the smooth running of an interview with a human to be interruptions, delays, perhaps incidents which spoiled the flow of a narrative and made the experience disjointed for my human interlocutor. On reflection, I think it might have been more interesting to have been less anthropocentric and more dogocentric (to borrow from Marc Bekoff); to try and understand more of how dogs engaged (or did not) with the process. Did dogs understand this kind of walking or hanging out as a process at all? As Lane remarks: 'I am unable to get inside Torridan's [her dog's] mind but I feel confident in suggesting she has no idea she is my research assistant—a walk is a walk'. It was certainly the case that there is inequality embedded in my processes and in my findings, for I prioritised one side of the human-dog relationship in this research; I did not treat all my participants with equal seriousness (Birke and Hockenhull 2015, 15–16).

My 'walk and talk' interviewing remains both a human and a human-centred method. It was the human participants who were accompanied on their usual route. How much more might I have gained had we let the dogs decide the route? Ironically, it may be that dogs exercised more agency when interviews took place in the home. My research assistants in the field had very little say, ultimately. Interviews in the space of home still featured interruptions—barking dogs, or dogs otherwise influencing the space (sitting on people or furniture, farting, nosily dreaming, attempting to 'steal' human food or drink, chewing furniture or clothing and so on) and, in turn, this led to shifts in topics and foci of human conversation.

In perhaps the most interesting case, within the first ten minutes of my first interview, Kevin actively participated in the emergence of a dog fight, or perhaps skirmish would be a more accurate descriptor. I am walking with two women and their four lurchers[7] when we encounter a man coming towards us with a border collie:

> Interviewee 1: Rory's ten or eleven this year . . . oops, collie alert, grab him [to partner]. They hate them [border collies] with a passion [to interviewer].
>
> Interviewee 2: They all hate them and if he goes [pointing to Rory], then they'll all go.
>
> [Before we can put the dogs on leads, the border collie runs up to us, fast, snarling with teeth bared. Kevin is misbehaving and winding it up—barking, snarling. Interviewee 1 loses her grip on the collar she is holding, and Rory goes to attack the collie, 'Get here boy', 'Rory, no!' 'Away', 'Get away now, come on' and so on. Kevin and Rory get inspected for damage.]
>
> Interviewee 1: What's that [on Kevin]? Is it mud or blood?
>
> Erika: Mud, I think it's just mud, that's OK.
>
> Interviewee 1: Off to a good start [laughing]! What happened there?
>
> Interviewee 2: It's Rory, naughty dog, Rory.
>
> Erika [of Kevin]: It's him, he started it!
>
> Interviewee 1 [of Rory]: Picture of innocence, look at him now [laughing]. It's collies, always a collie.

My initial reaction on completing the interview was disappointment that all had not gone smoothly and relief that none of the dogs had been hurt. In honesty, I would have to admit to being embarrassed that Kevin had failed in his role as my research assistant! Yet Kevin and Rory's intervention impacted data collection. In the immediate aftermath of this event, we discussed what happens when dogs dislike other dogs (on grounds of breed, or of personality), how this shapes the patterns of packs of human and dog walkers you socialise with, how it influences times and routes of walking, the experience of walking, the relationship with particular dogs and so on. I had not thought to ask about such matters, yet every subsequent interview contained discussion of varying lengths about cross-pack interactions, and this became a particular theme in the data on dog walking as generative of a particular kind of community, the subject of the following chapter. While Birke and Hockenhull are right to say that we cannot interview nonhuman animals for their view on relations with humans, in this case I think we did have a glimpse of what might 'matter to the animal from his or her point of view' (2012, 19).

ETHICS

> Interviewee [to dog]: Jago, did you give your consent?
> [To Erika] Do you want her paw print?[8]

While I 'went along' with those who cared very much for the dogs whose lives they shared in this project, I was always very aware that the domination of nonhuman animals shapes the lives of companions in dog-land, and the lives of dogs are precarious as we saw at the start of this book. In the stories of the lives of dogs and their human companions that I listened to, there are histories of neglect and of cruelty. For my human interviewees, however, upsetting stories of the past lives of the dogs-of-their-hearts, or of other dogs, were told willingly and people clearly felt that I needed to know such information. There was one question which I initially attempted to ask in interviews but quickly moved away from, and this concerned the loss of past dogs, the possibility of losing a current dog and the possibility of a future without a dog (due to age, illness or infirmity). People quickly became visibly emotional, and I considered another kind of project entirely would be needed in order to examine grief. The stories of the loss of a dog in chapter 5 are unsolicited but emerged in the course of a conversation; this was not a matter I could ask directly about.

In thinking about the ethics and politics of multispecies studies, Thom van Dooren, Eben Kirksey and Ursula Münster (2016, 16) consider that key questions 'follow from asking cui bono: who benefits when species meet?' Such questions also apply to the ethics of my study—to what extent did I genuinely include nonhuman creatures; and to what extent was I able to acknowledge and challenge the asymmetries of power between dogs and owners? In this extract from the field notes, I encounter four people, all of whom I have interviewed prior to this chance meeting on a routine walk:

> they ask how it goes. I say I've been interviewing quite a few people, and they ask Kevin how he is coping with his new role as my research assistant and all the walking. I say he had to walk four hours the other day. 'Poor Kevin, are you tired? She's terrible . . . etc'. 'You'll come in one day and find him on the phone to the Dogs Trust [UK charity]' and, impersonating Kevin: 'I've got this new job, but I never realised how much was involved. I'm an exploited working dog. Look at the length of my legs, they're worn away!'

While this was said in jest, Kevin was aging when he was involved with this project. That he was willing to be involved is without doubt—this dog loved to walk and always wanted to be where I was—but whether I should have planned more carefully *is* an ethical issue. While the dogs I walked with seemed mostly to be enjoying themselves, my presence was disruptive of routines in ways I had not expected and caused unanticipated anxieties:

> Erika: Where's Del gone?
>
> Interviewee: I don't know. Let's get a trot on and I'm sure he'll be up ahead eating something disgusting.
>
> Interviewee: If I'm taking to somebody and not concentrating, sometimes he will go off. If I'm focusing on him, I can preempt what he's doing. I'm not panicking yet. Although, he did run off and fell in [the river] between a narrow boat and the side of the canal, covered in green algae with just his little head poking out. If another boat had passed and washed it [the boat] in, there would have been nothing to be done. I had to do this commando roll and grab him by his collar. He's always sneaking off, looking for food, he knows there might be food on the boats he can get at. [We are running now.] Bloody dog! Bloody hell, I *am* starting to panic. Poo, human poo, there's quite a lot of that in these woods, so we're going back in there. I'm just going through the process of where he would go to. Shame you're not making a documentary, Erika; How are you going to describe this excitement and my fear?

'What am I doing?' I thought to myself, running through a wooded area with my concerned human interviewee, panting into the microphone which is poking out of my top pocket. We ask everyone we see; a woman I know on a large bright pink bicycle is alerted and volunteers to make a circuit around the woods where Del is found, predictably, 'eating something disgusting'. What am I doing, disrupting the routine practices of *this* woman and *this particular* dog, a dog who was so emaciated when he was found wandering the streets and taken to a shelter that years later, he still seeks out and eats whatever he can?

Apart from this one experience, I did feel that the human interviewees enjoyed their experience. During one interview, we come across another person who was soon to be interviewed. This prompts my interviewee to say to them, 'You are going to love this, it's great. It's just lovely just being able to talk about your dog to someone who wants to listen!' This, however, raises another ethical question that is associated with mobile ethnography—what to do about those you meet and speak with who are not being interviewed. Particularly instructive was another incident from my first interview, one which comes with a swear alert warning:

Interviewee 1: Oh Christ, look who it is!

Interviewee 2: Oh God, we'll be stuck.

Erika [laughing]: How long do you think it will take me when I get talking to [name of the elderly man and his elderly dog who are walking towards us]?

Interviewee 1: She has to do it [cut the conversation short], I just can't go, I can't do it. I'm pathetic, it's just what I'm like.

[Kevin barking.]

Elderly Man Who Is Not Being Interviewed: What's up Kevin? Come here, you come with me [strokes Kevin] . . . I'm going to report my neighbour next door . . . they've got a cat and it's beautiful, half Persian, ginger and white. But she never lets it in the fucking house. It's a beautiful cat. It comes up to me in the garden, but I can't take it in with 'er [his elderly whippet], she'll fucking shred it. I went to bed last night, I heard a noise, and I looked out the bedroom window and it's running around my shed. . . . And that poor little cat was still in the garden at half seven when I went out there this morning. I'll report it to the RSPCA. I don't mind a cat being out, but I mean this morning it was fucking cold, it was freezing. I got some food on the way home yesterday and put a saucer down, 'cause it was hungry, but I daren't take it indoors with 'er [his dog]. I don't like to see that. I mean, I'm not a cat lover, don't get me wrong, I *hate* cats, but it is a *beautiful* cat, beautiful, beautiful, it's only a youngster, about ten months old. I'm going to nick [steal] it, put a blanket over it, must be someone who wants it. Do you want it? I'll get my mate's carrier and I'll bring it to ya.

Erika: Only if it's an emergency.

Elderly Man Who Is Not Being Interviewed: It *is* a fucking emergency! It's got to be found a home; it's *got* to be found a home. It's a beautiful cat, tail's about that thick [makes a circle with his thumb and middle finger to illustrate] and right down to the floor. Half Persian. Honestly, Erika, I can't see that, I can't see that happen. I'm going to get some more food on my way home.

[Titus the whippet-collie appears, covered in shit]

Interviewee 1: Bloody hell, he's just had a bath as well.

> Elderly Man Who Is Not Being Interviewed [screwing up his face at the smell of Titus]: Fucking hell! I'm going—he stinks. [To Erika, with a wave of his walking stick] Find that fucking cat a home or I'll leave it on your doorstep!

I never was able to interview this man who was 'not being interviewed'. He became ill shortly after this encounter and, soon afterwards, died. He had been raised in Hackney and walked on Walthamstow marshes since he was a child. These recorded words encapsulated him perfectly—the phenomenal swearing, the rudeness, humour and kindness. When you research your own community, you document place, people and relationships. Would this man who was not being interviewed want to be documented? I suspect he would laugh and would certainly swear at the thought of being quoted in a book. I include this vignette as a form of memorialisation and do consider this as an ethical duty. In early publications from this project, I was concerned to further protect my anonymised human interviewees by anonymising their dogs (Cudworth 2011a; 2011b; 2017a; 2017b). The fast-paced life of a dog is short, and many canine members of the transient micro-communities of dog walking are no longer able to scratch and sniff. So, they are present here also, names and all, in most cases.

BEYOND HUMANCENTRISM?

In reflecting on my research into people's relations with dog companions in this book, I hope to reinforce the importance of an ethnographic approach in capturing nuance in human relations with nonhuman animals by inscribing daily lives and practices. This chapter has raised some questions about the inevitable humancentrism of ethnographic imaginations and the difficulties of bringing animals to the fore in our projects. Ethnography is a useful tool in attempting to map patterns of relationships and the ways lives are co-constituted in everyday practice. When I collected the data discussed in this book, I considered that I was engaged in research which brought dogs into the analysis, but the more I reflect on the material and how I obtained it, the more convinced I am that my practice remained humancentred. While I remain persuaded that 'walk and talk' methods were highly suited to this project, human talk and ways of seeing is inevitably privileged. My own research practice owed too much to anthropology and too little to ethology, and I see cross-disciplinary practices as vital in decentring the 'anthropos' from our 'ology', the human from our methods.

I do believe that telling a good story about co-constituted lives, that is both at the same time thick *and* critical on the page, is crucial in illuminating our reality in the more-than-human social worlds that we have a responsibility to represent. This is not just a question of 'science' but of politics and ethics. Not everyone might be interested in human-animal relations or multispecies scapes, but it is high time that the social sciences reflected on their anthropocentrism and positively embraced the notion of 'the social' as a complex network of multispecies lives, entanglements, dependencies and formations of power. This is an incredibly challenging task. For those of us who see ourselves as multispecies scholars, we are almost bound to fail. Certainly, the ethical, political and practical problems I encountered with my study meant it was far less successful than it might have been. There are ethical questions which deserve further discussion, such as the ways in which university research protocols need address the inclusion of nonhuman animals as research participants in a range of disciplinary contexts (rather than tools of scientific experimentation). What would it actually mean for a dog to give their consent to be involved in this kind of research? Are dogs vulnerable subjects in that they cannot give consent in ways traditionally understood? Alternatively, might we need to revise our understanding of how we register and monitor consent in the research process? If dogs are lively agents in the social world (and I firmly believe that they are), then how can this be acknowledged and theorised? Can dogs, as Lane asks, 'elicit stories that may not otherwise be told' by humans in our encounters? (2015, 27).

I was conservative in many respects in my approach and think that dexterous and cross-disciplinary collaborations offer more interesting future possibilities. Nevertheless, mobile ethnography can illuminate situated practice and document nuance in everyday lives. The chapters which follow are organised in terms of such practice and consider aspects of life shared with companion dogs in negotiating the space of home together, with respect to food and eating, the affective positioning of dogs as household members, and the impact of human work on, and the kinds of carework undertaken in, multispecies homes. Next, however, we will venture outside the home to consider the relational and spatial landscape in which people walk with dogs.

Chapter 2

Walking Edgelands

As gregarious, sociable and intelligent animals, dogs' need for different experiences and stimulation and exercising by walking with owners is common. This is often seen, by animal welfare agencies and people living with dogs, to be necessary and a key element of being a good dog owner. For dog companions in Western countries, being walked is the most common way in which dogs are able to engage with the world outside of their homes, on a regular basis. Dog walking is more than exercise for dogs and those who walk with them, however; it is a social phenomenon involving interspecies interaction and the generation of social relationships. This chapter suggests that dog walking is a timelised and spatialised practice where, in regular encounters over months and years, dogs and their human companions form relationships of community that might be understood as 'posthuman'. Such communities, this chapter will suggest, are loosely knit, shifting and tolerant of the diversity of their memberships. Posthuman communities are also distinctly located. The routine practice of walking through a particular space leads dog walkers to a knowledge of the places through which they walk and to the development of practices of care for those spaces and the creatures they find there—including other humans and their dogs.

Previous research has indicated that people who have dogs as companions are more likely to interact with others in public spaces, with dogs acting as social lubricants and encouraging human participation in community. This, in my view, underestimates the agency of dogs as productive in the generation of communities. This chapter will suggest that posthuman communities involves what Haraway (2008) calls 'cross-pack' relations—those formed outside of the home between humans and dogs when species meet through walking together. It will examine forms of intraspecies and trans-species companionship and conviviality which emerge in public spaces through the everyday practices of walking. Commencing with a discussion of concepts of community and posthumanism, this chapter also considers the location of posthuman communities in public places that might be described as *edgeland* landscapes, not fully urban or rural and often relatively unknown and uncelebrated. This conceptual landscape is illustrated and developed by drawing on my ethnographic analysis.

INTERSPECIES DWELLING AND POSTHUMAN COMMUNITIES

Those with rabbits, cats, parrots and other animals kept as companions in the home may engage in social networks in relation to the breeding, showing or rescue and rehoming of animals. However, dogs are distinctive in bringing their owners into public space through the need to be exercised (Wells 2004). While there is evidence that the ownership of a dog is accompanied by increased physical activity (Dembicki and Anderson 1996; Serpell 1991), in a humancentred world, where their lives are overwhelmingly shaped by their 'owners', not all dogs are exercised sufficiently. Quantitative research undertaken in New South Wales, Australia, showed that although dog owners might walk eighteen minutes per week more than non-dog owners, more than half of the dog owners did not walk their dogs at all (Bauman et al. 2001). Almost a quarter of dog owners surveyed in Perth, Australia, did not walk their dog (Cutt, Giles-Corti and Knuiman 2008). In the United States, a study of dog owners in Tennessee found that only 36 per cent walked their dogs (Thorpe et al. 2006); while a survey in Michigan revealed that only 27 per cent walked their dog for at least 150 minutes per week (Reeves et al. 2011). The Kennel Club of Great Britain[1] recommends different walking times for different breeds (more than two hours a day for some) but suggests a minimum of thirty minutes per day for all breeds and age of dog (Kennel Club 2013), a total of 210 minutes per week. A Kennel Club survey found, however, that one in five dog owners in the UK do not walk their dogs on a daily basis (Derbyshire 2010).

For the four in five dog owners who do regularly walk their dogs in the UK, qualitative research indicates that dogs motivate people to walk. A study conducted in popular walking sites in Hampshire found that dog walkers did not always want to walk, especially when weather conditions were poor, but that they did so out because they felt they had a responsibility to their dogs (Knight and Edwards 2008). Those regularly walking dogs (particularly where dogs are mostly off-lead[2]) demonstrate high levels of connectivity between different households (Westgarth et al. 2009). This is perhaps because dogs behave differently when off-lead—when dogs encounter one another on walks, they are far less likely to interact if both, or even one of them is on a leash (Westgarth et al. 2010). This chapter focuses on such dog walkers and argues that those who walk with dogs as a routine practice—despite the vagaries of the weather, their state of health, their work and other commitments—become, through regular encounters, communities. Unlike work which has been concerned with quantitative mapping of patterns of household connectivity, I am interested in teasing out the particular qualities of community in this context.

In thinking about more-than-human community, the notion of 'dwelling in mixed communities' of species is a useful starting point. This idea was developed by Norwegian philosopher and foundational figure in deep ecologism, Arne Naess (1979), who explored conceptions of community and its boundaries for humans who dwell in close proximity to wild animals. His emphasis is on 'self-realisation' that might enable us to (re)consider our notion of self and our practices as members of mixed species communities. Naess is concerned to de-center human interests and priorities, arguing for an ethics of tolerance and respect towards wild creatures perceived as a threat to humans, such as bears and wolves, and to question the uncritical defence of human interests and protection of domesticate animals (in this case, sheep) that are defined as human resources. The focus on relations between humans and wild animals may be a result of Naess's location or his deep ecologism in which domesticated animals are often considered uninteresting—bred to docility, dependence and 'stupidity', in contrast to the 'authentic' being of wild animals (Callicot 1980). While the notion of dwelling in mixed communities is an interesting one, my preoccupation is with dogs, who, as we have seen, are 'perhaps the only fully domesticated species' (Masson 1998, 29).

In a country such as the UK, as we saw at the start of this book, many of us already dwell in a mixed community of species within our own households. Work on the more-than-human qualities of the household (Fox 2008) has demonstrated how our ideas of family and home are often posthumanist. The terms 'posthuman', 'posthumanism' and 'posthumanist' are contested and deployed in various ways (Miah 2007). In my work with Steve Hobden, we have considered that the term posthumanism has been used in three principal ways: in the sense of a world after humanity (as seen in apocalyptic films, for example); as forms of body modification and transhumanist 'uplift' (where most often technologies are deployed to enhance human capabilities and release humans from bodily constraints); and in the sense of a world comprised of the more-than-human (Cudworth and Hobden 2018, 5). The sense in which they are used in this chapter and this book is in terms of the latter—a problematising of the fast distinction between the human and other species, and to denote an engagement with the challenge of sharing this planet with nonhuman subjects (Cudworth 2011a, 13; Wolfe 2009). Posthumanist understandings of the world resist what is often described as human exceptionalism, succinctly defined by Haraway (2008, 11) as 'the premise that humanity alone is not a spatial and temporal web of interspecies dependencies'. Hence a prime feature of posthumanist work has been to de-centre the human. This does not mean removing the human from our research, but rather to reduce the focus on the human and to pay attention to the web of more-than-human actors within which human activity occurs.

In this chapter, and the chapters which follow, there are moves across the human-dog divide, and there are spaces in which the distinction is blurred. In walking with dogs, other species are encountered and there are elements of dog walking practice which involve the exercise of care and concern beyond the individual human-dog relationship, and for species other than humans and dogs and in which multispecies intra-action can be found. People walking with dogs link with other 'packs' of humans and dogs, with other species found in dog walking space, and with the spaces, or, rather, the place itself. This chapter develops the idea of posthuman community by considering its distinctive features of eclecticism, dynamism, location and shifting constitution as found in daily practices of human and dog walking packs within a particular public space that might be considered edgelands.

THE EDGELANDS

The idea of the 'edgelands' comes from a paper by Marian Shoard (2002) which discusses the distinct features and contemporary significance of those spaces of 'interfacial land' emerging between many urban and rural landscapes. These familiar yet ignored spaces, that are neither town nor countryside, include business and retail parks on urban edges, disused quarries and mines, canals, ruins, piers and wasteland (Farley and Symmonds Roberts 2012). Urban parks are turfed for recreational football, municipally gardened and concreted and fenced for children's play, while the official wilderness of national parks is often overgrazed by sheep and heavily regulated. The edgelands, however, enjoy less pollution from farming and less management, and have often become rich in wildlife. In discussing the distinctive character of a variety of edgeland spaces, Paul Farley and Michael Symmonds Roberts argue that these force a critique of our romanticised pastoral of 'wilderness' and call for an acknowledgement of the wild places which are often very close to home. The chaotic wilds of the edgelands, they assert, has rather more authenticity that the pristine 'enshrined, ecologically arrested, controlled garden space' of much which passes for wilderness (2012, 8).

The dog walking communities of my ethnography are situated in edgeland contexts, but they differ from the sites explored by Farley and Symmonds Roberts. The dog walkers of a village in rural Leicestershire walk across countryside 'proper'. They navigate the problematics of horses, cattle and sheep, barns and wind turbines, hay fever–inducing crops, stiles, mud and puddles. They do so in fields crossed by official routes, that is, footpaths marked in the UK by yellow posts, and meander paths, or 'paths of desire', that emerge unplanned and fitfully over time from the favoured routes of various walkers (Farley and Symmonds Roberts 2012, 23–24, as can be seen in the illustrations in the previous chapter). Like urban dog walkers, they also walk pavements, albeit that theirs are thinner and at times disappear, and negotiating human litter is less of a chore. This constant movement along walks and paths, leaves limited room for interaction. In the places of congregation, however—by the hedges at the sides of fields or in small recreation grounds—there is more possibility for the emergence of posthuman community as people and dogs take time to gather and 'hang out' together. These spaces are examples of what Shoard calls 'interfacial land'.

The Lee Valley Park in East London is a rather different case of an edgeland space. The park is London's largest and follows the course of the River Lee (or Lea) from the town of Ware in Hertfordshire through parts of Essex and Northeast London to East India Docks basin on the River Thames. Parts of the park are designated sites of 'scientific interest' where rare plants or birds may be found, such as Walthamstow marshes which is a nature reserve.[3] Walthamstow is home to Europe's largest urban wetlands, and the park is a patchwork of more traditional marshland, such as the Lammas Meadow that has been 'upgraded' for public access, interspersed with areas of open space seen to hold less 'natural value'. An example here is Leyton Marshes, undergirded by rubble from the bombings of buildings in the Second World War and consequently well drained and well used. There are different sections of marshland along the river, and the areas used for this research covered Tottenham and Hackney Marshes, but was mainly concentrated on Walthamstow Marshes (towards the top right of figure 2.1) and Leyton Marshes (towards the bottom left of figure 2.1) and the Lammas Meadow which sits between them, all located between the River Lee itself and its flood relief channel (the river is at the top of figure 2.1, and the channel towards the bottom of the map).

The park is described by some of its users as 'the countryside', and there is an abundance of wildlife (rabbits, foxes, kestrels, woodpeckers, swans and voles on the river, and the occasional rare appearance of an adder or a cuckoo). While dog walkers have established routes along paths of planning and paths of desire, there is much space for meeting and stopping. In testament to its rather ambiguous status, however, some of the bridges are decorated with graffiti and an underpass had a mural painted by local artists (see figure 2.2); the marshes are cross-cut by rail lines and electricity pylons and parts are edged by housing and industrial estates. In addition, parts of the marshes may incur different temporary use from time to time—structures for training in the run up to the London 2012 Olympics, or a large marquee

Figure 2.1 An artist's map of Walthamstow and Leyton Marshes. Map credit: Lee Valley Park Authority.

to host the congregation of a local Pentecostal church, for example. They have also been used in the summer for ad hoc 'illegal raves', after which dog walkers encounter worse-for-wear party goers of a weekend morning in the summer months.[4]

Despite these differences, in both locations, the spaces of dog walking become an arena for interaction and gathering, examples of what sociologist Ray Oldenburg (1999, ix) calls 'third places' (that is, after home and work places) where unrelated people relate and communities 'hang out'.

GATHERING IN POSTHUMAN PLACE

When a space becomes a place, humans can be said to have affective ties to it (Tuan 1990, 4). Here, we return to the work of Yi-Fu Tuan, who had much to

Figure 2.2 Gypsy by the Mural on the Marsh on a railway underpass, painted by local artists.

say on space, place and their interrelation, and for whom the bond with place is constituted through our sensory perception as a species, alongside cultural perception and individual experience. Human sensory perception inevitably differs in range and acuity from that of dogs, as previously noted. While we are predominantly visual, they are olfactory driven animals. We nevertheless have much common experience in the spaces we share together. Both dogs and their humans are creatures of habit. Dog walking communities are timelised, and many of those humans and dogs who know each other well do so by virtue, at one level, simply of routine. Tuan considers that familiarity is key in the transformation of space into place (1977, 73). Our experiential knowledge of public spaces transforms them into places we have mapped and have a sense of intimacy with. This registering of a place is often developed says Tuan 'through experiences mostly fleeting and undramatic, repeated day after day and over the span of years' (183). Such repetitive and (mostly and *hopefully*!) undramatic experience and the process of familiarisation captures the qualities of walking with dogs well.

For Edward Relph (1976), our everyday and mundane practical knowledge of place leads not only to intimacy but to feelings of protection of places in which we dwell, and to the engendering of a sense of community. Oscar Newman (1972) argues that such intimacies encourage practices of care and a process of 'natural surveillance' through which space is unofficially minded. Whilst Newman is a controversial figure (as the architect of gated communities, see Harvey 1996, 292), this notion of intimacy generating a surveillance of place is an interesting one. As we will see, the ethnographic material suggests that the routine practice of walking through space leads dog walkers to a knowledge of the places through which they walk and to the development of practices of care for those spaces, and these include taking on roles which can be understood as a form of surveillance.

Michel de Certeau (1984) reverses the terms attached to these well-worn understandings of place and space, seeing space (rather than place) as what is lived, practiced, inhabited and remade on a daily basis. I am not convinced that's Certeau's reversal is particularly illuminating, but his idea of space (or conventionally, place) as reiterative is useful in thinking about the continuous constitution and reconstitution of the posthuman communities of dogs and dog walkers. Arturo Escobar (2001, 143) considers that 'place gathers things, thoughts and memories in particular configurations'. To this list I would add 'beings', for in the dynamic communities of dogs and dog walkers, new members, both human and dog, are incorporated and old members die, move away or otherwise leave and new configurations of community emerge, often shaped by memories of past members and past configurations.

There has been research which indicates that people with companion animals, particularly dogs, are more likely to interact with others in public spaces; and that such interactions are often facilitated by dogs (Messent 1983; Sanders 1999). Walking with dogs, for example, results in a significantly higher number of chance conversations with complete strangers than would be likely if walking alone (McNicholls and Collins 2000). In addition to enhancing social interactions amongst humans, dogs have been seen to act as catalysts for human reciprocity and aid (Guéguene and Ciccoti 2008), to facilitate community participation and to have a 'ripple effect' enhancing the sense of community in a locality beyond the relations between those walking dogs (Wood et al. 2007). This relegates dogs to tools enabling human interaction rather than actants in the production of community. Yet Deborah Wells (2004) has found that not any dog facilitates community—that younger and 'good tempered' dogs are stronger social lubricants than older animals or those which have received negative media attention (also Fridlund and MacDonald 1998).[5] Others have found that particular groups of dog breed (pastoral, working and terrier) seem associated with higher levels of canine and human interactivity (Westgarth et al. 2009). The notion that the interaction of dogs themselves facilitates community is important. Where dogs are able to interact freely off the lead, there is more likelihood of networks establishing themselves between households with dogs (Westgarth et al. 2009). The social interactions between humans and dogs, both as individuals (human-to-human or dog-to-dog contact) and packs, actually constitute community. This suggests that posthuman communities involve cross-pack relations in terms of intraspecies and trans-species conviviality.

In thinking about the particular characteristics of such communities, Oldenburg's work on informal public spaces of community is instructive. Oldenburg (1999) outlines key features of what he considers to be an ideal-typical setting for informal public life and compares examples of what he calls 'great good places'—including cafes and coffeehouses; bookshops, pubs, bars and taverns—to this ideal type. Oldenburg's concern is the decline of spaces in which informal community emerges with the development of suburban North America. He argues that such developments 'immunise' us from the development of community because they are:

> hostile to both walking and talking. In walking, people become part of their terrain; they meet others; they become custodians of their neighbourhoods. In talking, people get to know one another; they find and create their common interests and realize the collective abilities essential to community and democracy. (1999, xiv)

In the posthuman communities of the edgelands, walking and talking is the everyday activity and the mechanism for the emergence of relationships. Oldenburg's 'third' or 'great good' place is an informal gathering space that is inclusive and local. Such places have a number of characteristics: they unite neighbourhoods; assimilate newcomers; bring together those with similar interests; provide a 'staging area' for interaction; and enable the emergence of 'public figures' who know everyone and care deeply about their neighbourhood. Importantly, third spaces bring different generations together in the same space. In their convivial atmosphere, people get to know, to like and subsequently to care for one another; they also get to enjoy themselves 'daily and at little cost' and to get out and about. In addition, third spaces are political and intellectual fora where ideas are discussed, and where there is a divergence of experience and point of view (1999, xvii–xxvi). The discussion which follows uses elements of Oldenburg's ideal type of third space community in pinning down the specific qualities of the posthuman communities that dogs and dog walkers form, in addition to drawing on conceptions of attachments to place and the qualities of posthumanist encounters.

MAPPING POSTHUMAN COMMUNITY

Both the observational material and the interviews chart the practices of 'responsible' dog owners who walk their dogs very regularly, interact with them and have close bonds with them. Those who don't walk with dogs cannot be part of such communities, neither can those referred to by regular walkers in both the research locations as the 'fair-weather walkers', who walk when they feel like doing so rather than as a daily practice. Participants in this study already demonstrate practices of care for the dogs they live with by regularly exercising them. In addition, there are those who walk with dogs but who avoid contact with both other people and other dogs—perhaps because of their own aversion to sociality or that of their dogs'. This is not to suggest that all participants (either canine or primate) were highly sociable. Rather, different relationships emerge between different animals in their interactions, and while some quickly bond, for others it is more of a case of rubbing along together in edgeland space. The notion of 'dog people' as a community that is as much constituted by dogs as it is by their human companions will be discussed in terms of the themes of being 'outside', the ways dogs facilitate relationships (including friendship), the social diversity of posthuman community and the practices of care which emerge—the minding of place and the minding of animals (both human and not).

Walking and the 'Outside'

For Oldenburg (1999, 210–15) the routine practice of walking is vital for the creation of community. It connects people to public space rather than public space being something one drives through in the isolation of a car before arriving at a private home, which is the focus of leisure activities. There was a clearly established routine of dog walking at least once a day—usually more—in the lives of all my interviewees. For example:

> Walk him first thing in the morning about eight, for . . . a long walk of about an hour. I walk him in the afternoon for about half an hour, and then we have a quick skip round the block about, uh, ten or eleven o'clock at night. (F, Leicestershire)

> First walk at half past six, and I like to walk then. . . . And, um, then I'll take them on a proper walk, about half nine, quarter to ten, and then I go again just after lunch, depends what I'm doing. Obviously, if it's hot, I don't go until the evening. (F, Leicestershire)

Walking, in all weathers and at the darkest and wettest times of the year was viewed positively by every interviewee. The routine of walking was a key structuring device in the day, and for many, this was not a matter for human-exclusive planning but something that 'you have to negotiate' with dogs. People often expressed guilt if they did not adhere to routine times of walking, and others picked up on dog cues for walking with a number asserting that their dogs could 'tell the time'. The times and places of walking, however, are not always a decision to be made by human walkers. For some, routes and times are subjects of negotiation, while for others, it is the dog that decides where to go and when:

> It has to be the right time, doesn't it [to partner]? Set walks. We can walk him for five hours over . . . well, anywhere. We took him down to the seaside didn't we, but he still wants to go, when he comes back, to where he always goes, that's just kind of how he is. He's quite controlling. (M, Leicestershire)

Other people seemed drawn to dogs because of a love of walking or the 'countryside':

> we like walking, so that's very good and you have to go out for a walk, so even if it is six o' clock in the morning which is the only time you can walk him, you've to get out and walk him which for me is absolutely fantastic. (F, London)

> I didn't kind of get Elmo with the idea of 'oh it'll give me access to the marshes' . . . but I knew it [having a dog] would give me access to, er, getting out every day and you know, being in a bit of countryside 'cause it never occurred to me that I'd walk him on the roads near I live. Yeah, he's definitely my kind of er passport to, er, greenery. (M, London)

For most of my interviewees, however, it is having a dog that forces people to walk. This is despite poor weather, for some, poor heath, the time constraints of work and the extra work involved in drying out and cleaning up both humans and dogs on wet days and after muddy walks. This push outside and into exercise and sometimes exploration is to be welcomed:

> especially after work, you know, it's getting away from the desk, and getting the fresh air, and this time of the year, the fields of the buttercups are lovely and I mean you've seen Max bouncing, it's just such a pleasure to see that and throwing the ball and hearing him run—he sounds like a horse! (F, Leicestershire)

> We go on a lot of walking holidays in this country, and we wouldn't go and see so much of the countryside if we didn't have a dog. We wouldn't spend hours traipsing around Cornwall or Wales. (F, London)

> you wouldn't get up to get out just for a walk very often even if it was a nice day. We all want some excuse to go out, we want a reason to get out and do something, some reason to get out into the country and, er, dogs, dogs provide that sort of reason. (M, London)

> We have been out in the forest till sort of eight o'clock on a summer evening and it's just lovely and we might not feel like it after work. You think, 'No, actually I want to flop', and then there you are and you think, 'Oh isn't this great?', you know, 'I'm glad we've done it'. (F, London)

> I honestly think that [name of his wife] wouldn't be able to walk as much as she does now if it wasn't for going out with the dogs. Seriously, 'cause she makes herself go—

> [Wife interjects] No, it really hurts me. But like yesterday, we didn't get back 'til twenty to ten but it was just a lovely day, it was just a really lovely slow walk. We really enjoyed it. It was really nice. Sat on a bench. (M and F, London)

Some said that walking the dog(s) was 'frequently the high point of my day', others liked the fact a dog took them out of the house and encouraged them to 'do more things' or 'find new places'. Dog walking was seen as a necessity: 'It doesn't matter if it's raining, or snowing or cold, I'll go out. And I wouldn't do that if I didn't have a dog' (F, London); although many also enjoyed 'bleak walks' in the winter. Importantly, however, it is the walking that enables public communication, and without a dog, people tend not to walk so much in public space:

> my sister lives next door. She hasn't got a dog. She's got two cats. She drives to work and drives back, and she never walks anywhere. She doesn't know anybody so I'm telling her what's going on in the village because she doesn't know, and I know because I've got a dog and, when you've got a dog, you speak to other people with dogs or other people who haven't got dogs want to stroke your dog, so you always talk to somebody. (F, Leicestershire)

> I think if we didn't have a dog, it would just change everything so much, wouldn't it? I mean, I wouldn't get my ten thousand steps in a day. What would I do then, walk around the field by myself? That would be a bit sad. Wouldn't see all those people and chit chat, hear all their problems, for an hour or so. (F, Leicestershire)

There were negatives to the walking but mostly these came in the form of concern that 'old dogs can restrict your walking quite a bit'. This led to anxieties over the exercise enjoyed by dog walkers and the quality of life of dogs. Some, particularly if they had both older and younger dogs, would take dogs out even when they were unable to walk or did not wish to (in backpacks, trolleys, dog buggies and so on) in order to continue the processes of both human and canine socialising and 'keep our pack together'. The nature of dog walking, however, results in precarious encounters, so that even people who come to know each other well may not see each other for days or even weeks. The times and routes are important. While some follow set routes and times and will almost invariably meet, for those with more flexible routes and routines, there is a chance element to their encounters, and some felt their ability to sustain relationships with other packs was inhibited by the lack of a regular route and times.

It has to be acknowledged that not all walks proceed smoothly and that there were some tensions. In one case, a dog regularly refused to walk:

> There are things that we might wanna do that we know Jago doesn't. So, we don't do them. So sometimes I think, 'Oh God, I wish she wasn't quite so unpredictable about whether she'd walk or not'. Sometimes you get all prepared and everything. We had friends with a baby and a little boy who was desperate to walk with her and we got down here and she did a pee, did a poo, walked round the car park and got back in [the car] again [laughing]. (F, London)

While the overwhelming majority of dog–human and dog–dog interactions documented in the diary data are positive, there are occasional problems. In some cases, people will tether their dogs in the presence of other dogs or people they do not get along with; or humans will attempt conversation when the dogs of either pack clearly wish to keep a distance. Generally, however, interviewees in this study thought that walking enables public communication, and that without a dog, people tend not to walk so much in public space. For Oldenburg (1999, 210–15), routine walking is vital for generating and sustaining the local community. In both study sites, people felt that knowledge of both the places and people of their locality was facilitated by the walking that they undertook specifically because they lived with a dog, and they acknowledged the dogs' active role in initiating encounters with other packs.

Talking Dog

All the participants in the study agreed that one of the best things about living with dogs and walking with them was the people that you meet. People felt that dogs facilitated contact with others not just by virtue of being taken out for a walk but by functioning as a social lubricant. This maybe particularly important for new arrivals to an area:

> I mean when I first walked here, I spoke to all the old timers, with my dog, and now they all come and kiss me, and they're about eighty . . . my husband says, 'Who's that?' [laughs] and it's amazing, you know, what you find out and who you know and . . . you fit in better when you've got a dog, and you get to know people when you move to a new area. (F, Leicestershire)

Georg Simmel (1972) contrasted association related to social roles with that of 'pure sociability' in which people simply associate for the pleasure of each other's company. The need to walk dogs as part of the responsibilities of a dog owner could be seen as a social role encouraging a form of association in public space, but the qualities of association involved in dog walking also invokes Simmel's pure sociability. Almost all the interviewees commented on meeting people through walking their dog(s), stopping and chatting while out walking with a dog, staying longer than they planned as a result, or diverting their route to accompany other walking packs. People noted that other people would talk to them for the sole reason that they are with a dog, with a number remarking that people (with or without dogs) will start conversations making the assumption that people with dogs will be sociable. This may be because dogs are seen to legitimise walking as well as facilitate conversation:

> You can't walk without a dog, can you? I do see people, but it's not the same, is it? There's no purpose . . . having a dog helps you talk to people, there's lots of people now that I know. (F, London)

> People do come up and talk to you, there's camaraderie with other dog people. People say 'good morning' to you if you've got a dog. They might just ignore you if you were sort of wandering around. (F, London)

Most contacts between walking packs are initiated by dogs off-leash, greeting either other dogs or humans. Important also is that dogs themselves interact and force pauses in walks, encourage walking together and extended interaction. It is such extended interaction which facilitates the possibility of developing human relationships:

> If you go for a walk in the park you may say 'hello' to passers-by or you may not, erm, but it rarely goes beyond 'hello' . . . but once you have a dog and the dogs start playing together then, you know, you get into conversation with people and that's just a very nice thing to be social outside with other people, erm, just to pass the time of day rather than just walking somewhere and not saying anything or just saying 'hello'. (F, London)

The idea that walking without a dog is seen as odd, or that it is experienced as purposeless was common. In some cases, people even found themselves borrowing dogs in order to walk. Alternatively, groups of dog walkers may integrate those who are dogless into their routine walking: 'She's not got a dog, but she wouldn't meet anybody or see anybody in the village at all, if she didn't come out walking with us' (F, Leicestershire). Talking about dogs and looking out for each other's dogs binds dog walkers and provides ways into conversation in the absence of anything else:

> What do we chat about when we see other dog walkers? You chat about dogs, you talk about dogs and what dogs get up to and this sort of thing. You could say it's an inexhaustible subject of conversation! Now, I don't know whether it's just we're so bereft of bloody things to talk about that we have to, have to talk about the dogs but, er, but you do keep finding yourself doing so [pause] animals are quite intriguing. (M, London)

> Obviously, you end up talking a lot about the dogs when you see them [other dog walkers] anyway, there's a common sort of theme, but when they [the dogs] were obviously younger and you have the same types of sort of problems that they're having and things that you share in training, I tried this, or that, the other, you know, that sort of thing's interesting. (F, Leicestershire)

'We' are not just packs of humans and dogs from separate households, defined by our routes and our times, but also communities of dogs-and-humans walking together. It is often also the case that people may spend time talking over a considerable length of time, yet have gleaned limited information about other humans:

> You know the dog's name, but you don't know the couples' name. It's like them with, um, Ringo, we've known them for years and *years*, but we don't know their names. (M, London)

In many cases, even when names are known, the name of a dog is added for clarification, or their description becomes a pack term: 'the women with the Airedale', 'the woman with the Chihuahua, the staffy with the muzzle and two old ones in a pram' and 'the bloke with the cap and the dog with a tail that goes straight up like that' (accompanied by hand demonstration of the dog's tail).

All the interviewees did acknowledge that the groups one becomes a part of are strongly influenced by the dogs—if the dogs do not get along well, then people are rarely likely to exchange more than a passing greeting:

> There's regulars you talk to more than others, but then it depends on how your dogs get on as well. If the dogs don't get on, then you tend not to stop, so that sort of dictates who you speak to, really. (F, London)

> You know whose dog doesn't like who. There are some people that I would avoid, you know, some dogs that I don't particularly like and I would avoid if I could, but that's the same with people and people's children, I think. If you don't like people, you stay away from them. (F, Leicestershire)

Alternatively, close friendships were made as a result of the friendships established between dogs. All the dog walkers in the study had something to say about dog walking and sociality, and the often-close relations that develop between dog-walking humans, wherein friends are sometimes described as 'dog friends':

> It is an absolutely excellent amazing community and you meet so many people . . . and I have met, I am friends with so many people through walking the dog. (F, London)

> I'm a self-employed builder and apart from chatting to the clients I don't actually get to chat to a lot of people outside the family and that was an unexpected nice thing about having a dog, I got to meet lots of interesting people . . . I've made a couple of friends, in fact, in owning a dog and having a regular walk. (M, London)

> at the caravan site, you know, there's a lady and she came up t'me—she's quite a chatty person, you know—and she goes, 'I've lost my dog'. . . . You seem to pal up with each other, don't you? You've got something in common. Because she knew I'd lost Dennis [name of dog] and she said, you know, 'People don't realise, do they, unless they've had 'em' . . . it's only a site of about twelve caravans, and it's the people with dogs that stop, come to you and talk to you a lot, and you do them. (F, Leicestershire)

Certainly, the friendships of dogs are often compared by those who have children, to the relationships one makes because (or perhaps in spite) of them:

> It's opened up a new world of people and friends, really, people that you chat to, people that you see . . . she's [the dog] got her friends that she knew when she was a puppy, so they become your friends, like having children, with their parents, so there's lots of people that I know with dogs the same age that we met when they were puppies and all end up walking and meeting and chatting. The dogs seem to get on with certain dogs and not others, and then you get on with the owners. (F, Leicestershire)

There is conviviality which develops from the routine encounters, and opportunities for socialisation of both humans and dogs that emerge in and beyond the spaces of the edgelands. For Oldenburg (1999, 7) the 'problem of place' is an increasing lack of public places where people might anticipate unplanned meetings, where an informal public life might be generated. For him, third places are characterised by loose and unpredictable timings, and the charm of the third-place encounters lies in such chance engagements. The following excerpt from the diary data gives a flavour of the ad hoc social scenarios which emerge:

> Back by the horseshoe wood, [names of two dog walkers] are with friends, drinking and snacking to celebrate someone's birthday. I stop for a cup of warm white wine, Gypsy chases the boys around after some bread and Kevin is treated to bits of bread dipped in taramasalata, provided by little girls who make a great fuss of him. (Field notes)

While the timelised and spatialised qualities of dog walking as a practice mean regular encounters, there is an ad hoc quality to many meetings. It is a minority of walks that are planned with other 'packs', and these are usually between walkers who have become friends, or who meet for company at otherwise antisocial times:

> We always meet up with [names of dog walkers and dog] on a Sunday morning, and we meet up for dinner and stuff with them too. (F, London)

> I arrange to get here in the mornings at the same time as [name of dog walker]. It's good to have someone to go round with when you're early, especially in winter when it's so dark. (F, London)

> In the half-light, I see [names of two dog walkers and their dogs] at a distance, deep in conversation, crossing Leyton Marsh with thermos mugs of coffee. (Field notes)

Those with dogs that have been depicted in the media as problematic or potentially aggressive sometimes found it difficult to socialise with other dog owners and felt that people avoided them. For these walkers, the community of dog walkers is exclusive and prey to popular stereotypes of dangerous dogs. For others, the behaviour of their dog and anticipation of people's reaction to it led to a form of self-exclusion, as for this woman:

> So, the personality of the dog makes that social interaction. He [her dog] stops me from interacting with other dogs. Definitely. I'm very wary, my eyes are on fields all around me when Max is walking in case I see somebody and then have to put him on the lead. It depends on the dog owner. It very much depends. I mean, even this morning, I met somebody and because we were in this great big field, I thought I'm going to relax, I'm not going to panic, I'm going to let him do what he needs to do as a dog and this other dog was off the lead . . . so we let them play and, again, Max is boisterous and I apologised for him, as you do,

> and she said, 'Oh I think we've met Max before,' and straight away I thought, oh thank goodness, she remembers him, it's OK. (F, Leicestershire)

There was, however, concern that some of those who excluded themselves and their dogs from socialising did so because they were over-disciplining their dog. The vast majority of my interviewees had liberal views on training and all were convinced of the needs of dogs to socialise. Most expressed the need for dogs to 'be themselves' on a walk and to do the things they want to do, within the boundaries of safety for the dog and comfort of other dogs and walkers. The idea that some walkers control their dogs excessively when walking was common, and captured here by an older dog walker with experience of walking many dogs over decades:

> I see people constantly saying, 'Sit', 'Stay', they go on and on and on, I think, for goodness sake, they're a dog, let them be, leave them alone. There's somebody I pass—perhaps it's the dog, I don't know—but it's never off the lead, he's the only one in the field, and the dog is still on a lead, and I think no dog is that precious. I know some of them cost thousands, but they're not that precious that they can't be free. They've got to be free and have that run and sniff and do all the things that dogs do . . . I could quite easily say to the person, 'For goodness sake, let them be a dog first, they're not there for you to have someone to control'. (F, Leicestershire)

The relations with other dogs and their walkers stretch commonly held and sociologically well-worn notions of community as being about either place, or association and common interest. Dog walkers exemplify both of these elements. People walking with their dogs spend considerable time, over weeks, months and years, talking to other dog walkers. The communities of which one finds oneself a part depends on time and place—on the route one takes and the times of walking. Different communities emerge in different edgeland locations and at different times of the day, where, as often noted in the ethnographic diary, there are 'the usual people in the usual places'. Many interviewees spoke of 'their people' as the particular groups with which they speak most often, and with which they may perhaps also socialise. Those humans which form the closest bonds are those not only where the dogs are friends but also where common interests emerge through regular conversation, and these are very often linked to notions of appropriate care for and relationships with dogs.

Diversity in the Edgelands

For Oldenburg, a diversity of people from all walks of life and the novelty and interest sparked in conversation between those who would be unlikely to meet is a key attraction of third spaces (1999, 45–47). The communities of association in dog walking brings together a diverse range of people and the presence of dogs provides a talking point for people who would otherwise not be likely to socialise:

> You get to know people you wouldn't usually hang around with, people completely outside your usual circle of friends, and there's not many ways in which you could easily do that. (F, London)

> You strike up conversations with people that necessarily wouldn't have been part of your world . . . this very strange underworld of Hackney that I would never have tapped into, [but] because I was a dog walker, I was let into their world. I was on the same level. People that would never have talked to me before felt incredibly relaxed, told me their life story. (F, London)

> He's the sort of person in the village that you wouldn't normally, you wouldn't have come across, you meet him, and he comes out with some, like, little line when you're out on the field. He's a character. (M, Leicestershire)

Oldenburg (1999, 24) considers that third places are places where social levelling takes place. They are inclusive environments which are open to all-comers and in which personal status is irrelevant. One interviewee confided that she was 'fascinated by what people actually do' in terms of employment and lifestyle, because the dress codes of dog walkers (as one put it, 'They're not very smart, are they?!') usually leaves little clue to their social status and employment.

Dog walking brings together, for example, an older, white, heterosexual, working-class couple with a middle-class, middle-aged lesbian couple who become firm friends. The difference of age is also accommodated in dog walking communities. Different generations will rub along relatively happily together, from early teens to octogenarians. This differs from the norms of public space in which 'youth' are understood as encroaching on what is rightfully adult public space (Sibley 1995a; Valentine 1996) and are often understood as a form of risk (Chadderton 2013). Unlikely personal interactions made through dog walking, and the social facilitation of canine companions, enables a sociability that overcomes potential differences:

> There's this very tall lady, well it's a man with blonde hair who dresses up as a lady . . . and [there are] all the old people, all these pensioners sitting there, talking to her as if she's a lady. It's a bloke! But no, they don't worry, they accept this person, and everyone just gets on and talks about their dogs and their history. (F, London)

Many of my interviewees expressed the view that people who live with dogs are 'more tolerant' of other people and more sociable. Some further suggested that the reason for this is the process of walking and talking with other people and their dogs in which 'you open your lives up to other people'.

In addition, walking with a dog enables inclusivity in terms of security in public spaces, although this is distinctly gendered. '[She] stops, nervous in the dark as usual, making sure she knows it's me' reads an entry from my field notes, and one which is repeated in similar form across various encounters with women walking dogs in the dark. An element that has shaped my work with dogs and their humans has been that the majority of dog walkers that I have seen, and that have spoken with, have been women, and that walking in public space raises issues of risk. Most of the comments about risk have been almost afterthoughts in the stories dog walkers tell, but there is evidence that many women fear violence and police their times of walking. Many will only walk with someone else; some will only walk alone in the light; others prefer the dark because 'in the dark you and your dog can hide'. For the early morning walkers taking dogs out before work, there is limited flexibility and there are times when women express fear of male violence and that fear is associated with 'a man without a dog':

> It was really early and still quite dark, and there was this man who . . . wasn't dressed for walking and he wasn't with a dog . . . then I realised he changed direction—he'd gone under the bridge away from me and then suddenly he was behind me and normally Ziggy [her dog] doesn't pay attention to people but he kept running around this man . . . I was glad I actually wore my wellies that morning because it meant I could go anywhere, and unless he wanted to get himself covered in muck, he couldn't, so I sort of cut back across that meadow to the car park. I was frightened. (F, London)

In the case below, two women from London had lost a dog when visiting Ashdown Forest in Sussex, and had searched for a number of hours until:

> nine o'clock at night in December [therefore very dark]. The two of us, sliding all over the place in the mud and we couldn't see anything. And we had a squeaky toy and two heavy-duty torches and we were going 'Maaudy Maaaaudy!' and sliding about, and the next thing, in the darkness, in the middle of nowhere, this man appeared
>
> [Partner interjects] Without a dog.
>
> So, I thought he'd been burying somebody and she thought he was out sheep-shagging [laughing]. It shows you where our minds are really [laughing]. It was 'orrible! He freaked us out.

Yet despite this context of occasional risk, walking with a dog is seen as an enabling activity that functions to reclaim public space. This is both for women, and as the final quote illustrates, for men also, who are aware of the anxieties induced by 'a man without a dog':

> I saw the marshes as a dangerous place really. Somewhere that you wouldn't go, not on your own. But now I have Millie I'm just here all the time. So, she's opened all this up to me, the park, the marshes, I'm out there! (F, London)

> It opens up lots of opportunities as well, you know, like things that you might like not like to do on your own and so like going off camping for a weekend or something, I just feel a bit, you know, better about it so even like going for a walk in the evening you feel a bit safer having a dog with you. (F, London)

> What am I trying to say . . . I think it's something about being a bloke on your own walking about . . . because of, erm, you know, the climate of fear that we live in that people probably wouldn't talk to you in that kind of fairly random way [but] they feel quite happy if you've got a dog. (M, London)

Here, the presence of dogs is agential in transforming spaces of potential risk, and perceptions of risk. While the presence of dogs mitigates feelings of risk experienced by women, the 'man *with* a dog' is a less risky proposition—both for himself and for others.

Minding Edgeland Spaces

Despite the ad hoc nature of dog walking communities, the familiarity with the place within which one walks leads, as Tuan suggests, to a sense of intimacy. Dog walkers share 'secrets' about place (for example, in the London field site, the location of wild orchids) or the historical development of the edgeland space and its changing qualities and use. Some know an area since childhood and have an awareness of place that others may not have:

> 'This was my playground as a child', he says, waving his hand around over the Lammas field. We talk about how much we love it, and Epping Forest, where they usually go at weekends and where, I say, 'I am always getting lost'. He doesn't. 'I know it so well', he says. 'Certain trees, so I know where I am'. (Field notes)

In addition to knowledge, the narratives of dog walkers recall Relph's notion of the defence and care of place enabled by intimacy, and Newman's concept of surveillance. Dog walkers are both seen, and see themselves, as contributing to the spaces through which they walk and to the local community more widely. A number of dog walkers pick up litter (and for some this includes excreta of other people's dogs), and this may be a daily ritual.

Walkers, both with and without dogs, see dog walkers as keeping space safe. This is particularly so in the winter months, when, for example, 'the dog walkers are the only people that use the marshes, and keep an eye out, and know what going on' (F, London). Walking in all weathers and across much of the day in parts of the Lee Valley Park means that activity in the park is monitored, and even reported as necessary. Stories of reports made to park authorities and to the police included organised dog fighting, injured wildlife, abandoned vehicles and even a case where dog walkers discovered a woman who had been sexually assaulted. This last case was particularly significant. In the immediate aftermath of the attack, it was dog walkers who called the police and looked after the woman until they arrived. While the police refused information on the incident to members of the public, dog walkers informed other walkers in the park over the following hours and days about this and further instances of sexual assault, and organised themselves into walking groups so that women would not need to be walking there alone. It was dog walkers who provided information leading to an arrest and prosecution. This notion of a permanent community of people who 'keep an eye out' is in keeping with Oldenburg's ideal of publicly spirited individuals in public spaces:

> well, I mean it's a nature reserve, there's the plants there that don't grow anywhere else but, if push came to shove, it would be all the dog walkers protecting that area down there . . . if it wasn't for the dog walkers, then I think that, y'unno, it would be a different place down there, I really do . . . the dog walkers make that a safe, accessible place for everyone to go whatever time of day. (F, London)

Communities of dog walkers are active both in preserving edgeland space and in promoting their interests as users of open spaces. In the year prior to the London 2012 Olympic Games, dog walkers were an important group within the membership of the Save Leyton Marsh campaign which lobbied and protested against the siting of a temporary olympic training facility and for the restoration of the site.[6] In response to plans from the London Borough of Waltham Forest to impose bans on dogs, small dog exercise areas and leash-only dog walking in some of the Boroughs parks in the spring of 2013, a campaign of protest resulted in a consultation exercise and the formation of a pressure group, 'Waltham Forest for Dogs' (Curtis 2013). In the Leicestershire village, concern about upsetting dog walkers and potential protests was given as a reason for alterations in plans for the development of a health centre, enabling two large, open fields to remain under pasture rather than being used for building and car parking. The notion that dog walkers engage politically in order to preserve the integrity of the edgeland places through which they walk, and the practices within these places, articulates Oldenburg's idea that third places can be those in which political ideas are flouted, engaged with, refined and have possible effect.

Minding Animals

> People will ask about other people, and if you haven't seen somebody you think, 'Oh god, where's so-n-so?'. Yeah, that is really quite a unique thing that happens. (F, London)

> People look out for each other. You hear through the dog walking grapevine. (F, Leicestershire)

Marc Bekoff's (2002) notion of 'minding animals' is that humans might act towards other species with both compassion and care, together with respect for the lifeways of nonhuman beings as 'minded animals'. For Oldenburg, the intimacy of routine and unplanned hanging out results in people having an awareness of how things are and should be and taking action to protect the ways of being, and the beings which inhabit, his 'great good' places. In this diary extract, a dog walker finds a dog out of place:

> 'Is that your dog?', he asks pointing to the black and white puppy. The woman says it has followed them. He says it belongs to someone on one of the narrow boats [on the River Lee] and he will take it back. The woman says she will go back that way so there is no need. It's all very amicable, but he is quietly persistent. . . . Later, passing a boat, a guy calls me over, anxious, had I seen a collie pup? I'm just explaining and then turn to see [name of dog walker] in the distance, walking fast towards us, waving, with the puppy on [his dog] Feedback's lead. (Field notes)

Taking time to stop and talk is characteristic of dog walking practice. While routes are often established and normative, people will shift direction so that they and their dogs might exchange more than a brief greeting. In some cases, the practice of care is simply the process of greeting and talking:

> A long and slow walk in the afternoon with [name of dog walker]. I moan about the weather, but he tells me how good it is. He likes the cold and doesn't mind the dark—as long as it doesn't rain. He likes the summer light but not the heat. It is dark and late when I return. Definitely worth the slowness in the cold—'thanks so much for taking the time to keep me company', he said, as we left each other. (Field notes)

This configuration of care and respect is present in attitudes to both dog walkers and their dogs, and is particularly evident in situations of loss. Dog walking communities are characterised by transience—dog people age, become ill and die or move house or location and enter another dog walking community. The relatively short life span of the dog means that people may slip out of dog walking communities temporarily or, on occasion, permanently after the loss of a dog. The interview and diary data document high levels of concern for the health and well-being of dogs amongst dog walkers, and that dog walkers appreciate the empathetic understanding they receive from other dog walkers when dogs are ill or lost. The diary data also documents the networks of people and processes gone through in a case of finding a home for a dog abandoned by its owner.

Oldenburg's third space communities are characterised by the presence of key public figures, and amongst dog walking communities, these can be both canine and human. In the calendar year when the ethnographic diary was kept, the dog walking community of the Lee Valley Park saw numerous dog deaths—some of aged dogs, some of ailing dogs and some due to accidents (such as anaphylactic shock from a wasp sting). It also lost some human 'public figures'. One of these was probably the best known of a group often referred to by other dog walkers as 'the old boys' who have spent all their lives associated with Walthamstow, Leyton and Hackney Marshes, had detailed knowledge of the places and their history and were a well-known sight. The death of this man, who we met in the previous chapter (not being interviewed!), is a narrative that runs across the diary data from his quick illness and death in January, until the last pages in December. The concern for this person, the grief expressed at his passing and the worries over the health and happiness of his dog in his absence are indications of the networks of care and concern for humans and dogs. They also illustrate the point Oldenburg makes about third spaces enabling the emergence of 'public figures' who know everyone and are known, who care deeply for and become strongly identified with place:

> 'Have you seen [name of dog walker]?', he asks. I tell them. He says, 'Oh god, no', and turns away. She says they've been phoning him, no reply. 'It won't be the same without [him]', he says. 'He was like a part of this place, if you know what I mean'. (Field notes)

In this and other cases, deaths, illnesses and moving away from the area lead to reconfigurations of walking patterns and of networks of people and dogs in the fluid way suggested by de Certeau. Yet also, as Escobar asserts, new spatial configurations are haunted by memories of people, of dogs and of the relationships between them.

POSTHUMANIST COMMUNITIES IN THE EDGELANDS

This chapter has argued that people and dogs, walking together in certain spaces that we can consider edgelands, might be considered a community. Theirs is a fractured community, perhaps best understood as a plurality of micro-communities that are defined by time (the regular points of walking) and space (specific routes in which dogs and dog walkers encounter one another). These communities are characterised by both a commonality of place, and attachments to it, and a commonality of interest in walking, in hanging out outside and, for the humans who have agency in deciding this matter, an interest in common in terms of their experience of lives with another species. Not all dogs and dog walkers are embedded in the communities discussed here. Some dogs and some people are antisocial and may self-exclude. There are divergences within packs on this matter—where either dogs or people may choose to join in where their pack mate selects abstinence. There is also some evidence of ostracism. This said, in both locations, there was evidence that most regular dog walking packs are connected in positive relationships to some degree.

Posthumanist communities of walking packs have certain characteristics. First, there is the peculiarity of mobility and the situation of being outside. While there is certainly identification with outside place(s) this is a shifting and dynamic location for the emergence of community, particularly as the community is generated, reproduced and intensified through the process of walking. Second, this community is not exclusively human. At one level, dog walking communities consist of dogs and their humans in individual and cross-pack interactions. In addition, the agency of dogs is apparent in the creation of interactions between dog walkers, and this is key to the deepening of relations or the absence of any kinds of connections. Both these elements are sufficient for such communities to be understood as posthumanist. Third, these posthuman communities are characterised by a relatively high degree of tolerance of diversity. Humans from various kinds of social backgrounds are brought into proximity through the process of dog walking. In addition, there is tolerance of some diversity of human-dog relationships and of dog personality and behaviour, but this is less pronounced. Finally, there are a range of practices of care and respect within these posthuman communities that exemplify the notion of minding animals. Within these communities there are high levels of concern for the health and well-being of both human and canine members and also care for place. At times, more traditional forms of political engagement emerge from the debates and discussion amongst human walkers and focus on the protection of edgeland space and walking access and the generation of 'dog-friendly' public policy. As such, I consider emergent communities in edgeland spaces to be heterotopic—suggesting an alternative idea of community as one that is at the same time affecting, while also being transitory and shifting in nature.

A question, perhaps, concerns the extent of cohesion of this community—whether the timelised and spatialised qualities of engagement mean that it might be best to understand packs of dogs and dog walkers as forming a series of micro-communities only. There is, however, evidence of links across micro-communities of packs and an overall impression, in both research locations, of a linked network of cross-pack groups. A strange assortment of people and dogs may be found interacting in the edgelands, forming, I would suggest, vibrant, eclectic communities of a distinctly posthumanist kind.

Chapter 3

Sharing Home

In this chapter, we move away from the public spaces of walking with dogs to the private space of the home. The key motif for this chapter is mud, or rather muddiness. One of my human participants claimed that 'It gets a bit mucky dunnit? Life is just muckier when you've got a dog. It's part of the deal. It's probably one of the good things about living with animals actually' (M, London). This chapter uses the idea of 'muddied living' in order to suggest that living with animals as companions in the home is messy in two ways, both of which are deeply material.

First, there is, as my participant suggests above, the issue of the extra 'stuff' that nonhuman animal companions—in particular, dogs and cats—bring into the space of domestic dwelling in the Global North. A while ago I emailed a friend in despair having undertaken a survey of my home, after another friend had come to stay with two dogs and an elderly cat. This meant two days of a house occupied by four cats and five dogs. Prior to the arrival of our guests, the house had been thoroughly cleaned, vacuumed and all bedding washed, yet my email noted the presence of:

- Cat hair: various. Red hair from aging visiting cat found on duvet, bed sheets, top of printer (he is more agile than we thought!), keyboard of computer, on top of a pile of books and on kitchen table. Black hair from resident cats on bottom of curtains and on every chair in one room, also on kitchen windowsills. Long hair on one chair only—Penny cat's favourite, which all other creatures are too scared to occupy. Also found a fur ball of long cat hair and accompanying vomit stain on door mat.
- Dog hair: various. Light coloured hair mostly visible, different shades for different dogs, on all bedding in house, most chairs and particularly on large sofa, as well as on dog beds.
- Mud: paw prints from cats (wet feet from rain/wet garden) on furniture and an array of bits (soil, moss, green and dried brown leaves, small bits of twig). Cat paw prints across floors of two rooms. A tennis ball covered

in mud had been brought inside and the felt partially ripped, possibly also eaten. Mud/grit found on sheets on beds in all bedrooms and traces on soft furniture. Mix of grit/grass around back door to garden.

- Urine: soaked towel sitting in washing machine waiting to be washed along with various cloths used for clearing up, possible missed bit—dried urine stain on tiles near back door and skirting board? Cat grit (from tray used by elderly visiting cat, and which had to accompany his movements) scattered on bedroom floor (stuck in grooves of floorboards), on top of bed and under duvet, two 'bits' stuck in keyboard of computer, in kitchen on floor, on mat, chair and four bits found in a shoe.
- Vomit: see above *re* cat hair fur ball, also appear to be dog vomit traces between paving slabs outside back door.
- Shit: of various consistencies across lawn and on flower beds.

Clearly, my household was, and still is, well exposed to a wide variety of microbial life. However, such an array of disruptions to household hygiene and tidiness are routine, to some extent, in the homes of all those who share their space with cats, dogs and other critterly companions. One morning at the time I first began this chapter, I walked into a crime scene in my kitchen—two dead mouse bodies, one of which had been decapitated. Serial killers Kali and Tinkerbell were unresponsive to questioning, but blood staining and initial olfactory investigation by Vinnie the Jack Russell suggested that the bodies had been dragged around the kitchen before being strategically positioned. Some disruptions are more disruptive than others. A day before a second drafting of this chapter, after two months, a petrified severed mouse head was found in the house, while yet another mouse corpse was presented outside the back door. Such horrors are not the sole preserve of cats—on a number of occasions, I have been unable to prize a dead rabbit's body from the jaws of a dog and such sad treasure has been brought back home from a walk. The comparatively greater freedom of cats, however, does raise challenges for those of us who seek more harmonious relations in the multispecies home, as Michelle Westerlaken so well illustrates (see figure 3.1).

Apart from the routine piss, shit, mud and hair, and the thankfully relatively rare corpse, there is another way in which living with animal companions profoundly muddies human lives. This is by calling into question the boundaries between humans and other home-dwelling creatures in terms of species boundaries and in terms of the use of space within the home. Commencing with a discussion of posthumanism in relation to the home, this chapter illustrates and develops the notion of muddied living drawing on the ethnography. Almost all the human interviewees made the point that the quality of home is altered by the presence of animal companions. Many suggested that it is a 'happy home' that has a dog, or a more 'homely home'. More radically, the

Figure 3.1 Illustration by Michelle Westerlaken, from *Imagining Multispecies Worlds* **(2020).**

presence of a dog is seen to transform a domestic space, with many uttering a variation of the phrase that 'a dog makes a house a home' because of either or both the kinds of specific relationships that muddy the clarity of human/nonhuman animal boundaries or the literal muddying of a multispecies domestic space. Such muddiness may, however, be far from the kind of 'ideal home' we encounter in the media and popular culture, and to set the contrast with posthumanist homemaking, let us first consider contemporary ideals of Western homespace.

IDEAL HOME?

Houses or other kind of dwelling space and the idea of home are often conflated in the popular media, typically as a means of selling property and promoting 'home' ownership. Neoliberal governments in relatively wealthy countries such as the UK have actively promoted the conflation of house, home and family as an element of an ideological agenda aimed at increasing economic growth and emphasising the significance of traditional notions of the family and the relative importance of the private sphere (Madigan et al. 1990; Sanders 1990). The relationship between house and home has often focused on the notion of the ideal home, one which is often influenced by the views of interior designers and shaped by cultural norms around the presentation of homespace, for example, as 'clean' (Chapman and Hockey 1999). The latter is particularly cultural significant, as Ruth Madigan and Moira Munro's (1996) work on the 'beautiful' home found, for while women were less certain on matters of appropriate style, there were clear pressures on them to produce homespace that was clean and tidy.

As feminist work long has argued, cleaning homes is arduous and necessarily repetitive (Oakley 1974), and as we will see in chapter 6, living in a multispecies home does come with costs in terms of domestic labour. Life, and bodies, are leaky. Dogs perspire from their paws, and from their hair follicles. Their sweat has a chemical scent that is individual to the dog, so while dogs may smell the same to some of us, they certainly do not smell the same to each other. They also produce oil, an important part of healthy skin and hair, which has its own scent, and have glands in their ears, which produce a light yeasty-smelling wax. These smells are all normal body odours for a healthy dog, but they cause horror for some humans, particularly when trying to sell or buy a home. *Cleanipedia* (sponsored by Unilever) paints dog smells rather differently—'large parts' of dog bodies produce sweat, their pads spread this around the home and dogs are 'constantly producing sent marking odours'.[1] *wikiHow* advises on how to eliminate such smells by bathing a dog, cleaning out its ears, cleaning its teeth and transforming its flatulence with home-prepared food.[2] Anxious questions are answered online: 'Should I wash my Yorkie [Yorkshire terrier] with bicarbonate of soda and vinegar?'. Thankfully, no.

Looking at internet advice on how to sell a home with resident animal companions, the notion of the attractive and 'ideal' home is far from the home with animal residents. While sellers of homes are represented as ignorant of the animal smells in their homes, the home buyer, on the other hand, is represented as having a highly acute, almost dog-like sense of smell, such that one website warned that potential buyers are able to smell an old stain on a carpet immediately as they walk into a house. As a result, the elimination of the negative equity wrought by living with a canine companion means that deception is required:

> If you want to ensure you get the highest final sale price, you'll have to properly stage your home to create the illusion you are not a pet owner. . . . Not only are odors very obvious when buyers first enter your home, they are usually very difficult to get rid of. . . . Where do these odors live? Everywhere. Try to make sure you know each location your pet had an accident. . . . Did your cat get sick on the windowsill? If so, a simple paint job to cover the stain won't remove the smell. You'll have to replace the wood to completely remove it. The same is true for flooring. If your puppy had an accident on the carpet, it's important to replace the carpet, as well as the pad below. And substantial stains can even seep into the subfloor through the carpet pad. If you think this is the case for your home, it's best to hire a professional if the subfloor needs replacing.[3]

The advice to the homeowner living with an animal companion is hardcore, particularly if it comes from US sites. Board the animals at another property while you sell your home, wash everything that can be washed, replace all furnishings that cannot be washed, replace or have your carpets professionally cleaned, scrub all hard surfaces with bleach, buy an ozone machine.

Given the discussion threads on how to keep homes clean, it seems clear that the extra cleaning smelly dogs necessitate is women's work. Mumsnet and Netmums sites have trails bemoaning elderly 'stinky but lovely' dogs and the difficulties of dealing with odour. Advice abounds: bowls of vinegar are to be positioned around the room, essential oil dripped onto tea bags and put inside the vacuum cleaner, 'Shake and Vac' products or bicarbonate of soda is to be used to restore freshness to carpets and furnishings, hard floors should be disinfected, a new vacuum cleaner may be needed. The critter itself may need bathing (at home or, better, professionally), wiping down or spraying with deodorant. Even the forum of *Horse and Hound* (the oldest, weekly equestrian magazine in the UK), which might be expected to be less concerned with household appearance, has worried readers asking for help. The owner of two Labrador retrievers, for example, while claiming that she 'loves' her dogs, seeks advice because the 'state' of her house is driving her 'completely insane':

> There is always dog hair EVERYWHERE—it sticks to everything and anything, and I can spend my life hoovering [vacuuming], only for them to be let out and cover everywhere again. The walls . . . are often covered in splotches of whatever they've brought in from outside. . . . They manage to soak the house by drinking water, then running through the kitchen and dribbling water as they go. This means the floor is always wet, so my socks are always wet, and the dog

> hair sticks to the wet so hoovering [vacuuming] doesn't pick it up, only moping, which effectively just spreads the hair across the floor. Our house always smells like dog, too.[4]

The advice she receives ranges from the usual bicarbonate of soda and related products, to purchasing expensive brands of vacuum cleaner designed for picking up animal hair, using an industrial dry steam cleaner and an 'industrial hot extraction carpet shampoo'. More considered is the UKs premier home design magazine, *Ideal Home*, and its associated website, which recommends, along with the expensive vacuum cleaner, replacing carpet with hard flooring, avoiding pale colour schemes while eschewing toxic cleaning products which may endanger dogs.[5] However, the hundreds of photographs of home interiors from 'ideal homes' to be found on their website indicate that the ideal home is without dogs. Because homespace is idealised, and a reflection of self and identity (Mallett 2004, 82), sharing homes with dogs compromises the idealised homespace to which bourgeois popular culture encourages those who can afford it to aspire.

POSTHUMANIST HOME

As we have already seen, the contested terminology of 'posthumanism' or 'posthumanist' has been deployed to describe various discourses and philosophical claims about the human, the animal, nature and artifice (see Braidotti 2013). Posthumanist scholarship is concerned with developing analytic frameworks that account for the more-than-human constitution of the social. This is not simply a case of demonstrating that the social world is constituted by nonhuman animals but of drawing our attention to the *co*-constitutive character of human/nonhuman lives and relations.

Work on such species encounters within the household has described the ways in which boundaries between humans and animals, friends and kin, are porous and shifting (Mason and Tipper 2008). Empirical studies indicate posthumanist relationships in the everyday practices of pet-keeping and a negotiation of human and nonhuman animal identities. For Rebekah Fox, it is the uniquely liminal position (2006, 526) of animal companions in the home that makes this interspecies identity work both possible and necessary. It is interesting to note that research has found that children far more readily closely identified with animal companions compared with adults in the same household (Tipper 2011). While the nature of the bond between people and animal companions may be seen as having different qualities to those between human members of the household (Walsh 2009), there are contradictions and tensions. For Fox (2006), animal companions are constructed as both family members and as almost human in their individuality, yet on other occasions, they are defined by their instincts and essential differences from humans as animals. Such ambivalence suggests that boundaries may be both maintained and breached in posthumanist households.

Nickie Charles rightly argues that close relations with animal companions are nothing new in the West, although levels of intimacy may have deepened through the twentieth century and beyond. As a result, she urges caution with the use of posthuman to describe a recently emerging formation of the household (Charles 2016, 2). She argues that we need a more granular quality to our thinking about relations with animal companions which allows for different levels of intimacy with specific individuals rather than a collective animal that serves as a companion (2016, 8–9). We also need to think about the role of the animal companions as actants in relationships—intimacy is 'constructed through the actions of that animal' (2016, 9), and is fundamentally inter-corporeal, with bodies sharing touch, contact and intimate spaces. I concur that the posthuman does not signal a new social form but, rather, is a recognition of social life as ever multispecies (Cudworth 2011; 2016a). As such, I am comfortable with the appellation 'posthumanist' to describe the multispecies home in order to capture that the quality of home is *altered* in its more-than-human constitution. This chapter is focused on a surprisingly marginalised sociological matter, that of home as a site of spatial relations in which humans and dogs interact. The construction of dogs as family or kin and the affective relations with the animals sharing the household will be the focus of chapter 5.

Sociologists have paid limited attention to the place that is 'home' and how it is used and shaped by occupants. Christopher Lasch's (1995, 3) well-known account of the family as the 'chief agency of socialization' which 'reproduces cultural patterns' ignored the spatial dynamics of home life. Yet domestic relations are spatialised, they are enacted somewhere. Such spaces are varied cross-culturally, shaped by intersected forms of advantage and disadvantage and vary in material form from a suburban house to a trailer or a street corner (Kusenbach and Paulsa 2013). For Thomas Gieryn (2000, 264–65), who has insisted on 'a place for space' in sociology, the place of home varies in the way it is invested with meaning and value. The sociology of home is best described as 'nascent'. This is surprising, given that so much of sociology is *about* home, from sociologies of the family, community or migration (Anderson, Moore and Suski 2016, 2–3) to those of aging and the lifecycle and emotion (Kusenback and Paulsa 2013, 2).

In developing a sociology of home, sociologists have drawn on the humanistic tradition in geography where the understanding of home as a haven or space of refuge in which we emotionally invest comes through strongly (Tuan 1977, 1996). This is reflected in the title of Lasch's work on the family as 'haven in a heartless world'. However, place—home or otherwise—is not fixed, permanent or rooted in notions of the authentic; rather, the constitution of space is always in the process of construction (Massey 2005, 9). Home is understood as a 'center of meaning and a field of care' (Cresswell 2004, 24) wherein people might be 'themselves' and as having strong claims on the time, resources and emotions of its occupants. Peter Sommerville (1992) suggests the meanings of home might be categorised as 'shelter' (physical security), 'hearth' (comfort), 'heart' (a site of love), 'privacy' (a boundaried space that its occupiers' control), roots (a source of identity), abode (a place of rest) and 'paradise' (an ideological construct fusing positive features of home). In trying to capture the extensive emotional ties conjured by the idea of home and place, some deploy the notion of 'belonging' (Savage, Bagnall and Longhurst 2005), while others reject this as effectively fixing home in a locality rather than understanding home as constantly in the making as people *em*place themselves (Korać 2009). Maja Korać uses the term 'nesting' to connote dynamic processes of making home (2009, 31; also, Ingold 2012). In nesting together, humans and dogs invoke a number of Sommerville's practices and meanings of home, in particular 'heart' (practices of intimacy, love and care) and 'privacy' (a space for the expression of self, although this is also a space that humans control). Sommerville's notion of 'paradise' is too strong, and I consider that the multispecies home might be better understood as a heterotopic space, a fractured, multiple, partial glimpse of utopia, at best (Foucault 1986). This is a matter to which I return in the conclusion to this book.

The home is also a matrix of heterogeneous social relations. Feminist scholarship has a legacy of critique of home as a privatised space in which oppressive and exploitative gendered practices and relations cohere and are relatively immune from public intervention. Gillian Rose (1993, 47–51) is rightly troubled by the universalism of Tuan and others, the lack of attendance to difference and the practices associated with it including reproductive labour, the costs of social reproduction and the danger of domestic violence (also, de Lauretis 1990; Hochschild 1997). This lack of attention to gender, says Rose, marks home as a masculinist place of desire (1993, 53). Not all feminists share such a view. The home can also be understood as a site of resistance, enabling a dignity and strength that Black women in the United States, for example, may have been denied in the public realm. The work of bell hooks (1990) is well known here, reflecting on the home as a space of care and relative freedom in a racially exclusionary and often threatening public world. For hooks, home is both a refuge, and a place of empowerment. Similarly, the home might be a site of political resistance and emotive support for lesbian (and other non-heteronormative households) in a heterosexist public world (Valentine 1993; Elwood 2000), while fear of public space may, for diverse women, render home a place of safety. In a humancentric public world, a question worth asking is whether the private space of home might also be a space of relative freedom for some dogs and their humans.

For many humanistic geographers, invested space such as 'home' is constituted reiteratively, made and remade through everyday practices. Each home makes its own world through the repetition of routines of sleeping, grooming, relaxing, cooking, eating and cleaning (Bowlby, Gregory and McKie 1997). The home is also a site of negotiation of spatial boundaries between household members. David Sibley (1995b) considers tensions between parents and children over the use of household space, which is both rule-bound and constantly negotiated. Children do not passively accept adult spatial boundaries within the home (Valentine 1999). Within everyday spaces, the social is constituted as heterogeneous, relational and importantly 'lively'—dynamic, shifting and emergent (Massey 2005, 13, 55). A spatial lens pushes us towards the more-than-human—the place in which we dwell, the objects and (in the case of animal companions) subjects in which we emotionally invest.

Like children, despite some communicative difficulties, dogs do communicate within the context of home. Sociologist Clinton Sanders kept field notes of his developing relationship with two young dog companions, noting that rather than simply talking to himself when alone at home, his 'self-speech' became increasingly directed towards the dogs (1999, 14–15). In his interviews with dog owners, Sanders found many accounts of people talking to dogs, most usually however, they did not consider this to be self-speech directed at animals but a situation of mutual communication between themselves and the dog(s) they lived with. Similarly, accounts of dogs 'joining in' with rituals within the home suggest mutuality rather than humancentred and directed behaviour (1999, 29–30). Owners of animal companions, particularly dogs, also 'share their meals with them, allow them to sleep on the bed, and to sit on the furniture, even to celebrate their birthdays' (Serpell 1996, 74).

Writing on animal companions in the household, however, has rarely emphasised the impact of animals on home as a space. Where it has, Jacqui Gabb (2008) has suggested that being seen as a close family member is reflected in the fact that private spaces within the Western domestic home (such as bedrooms and bathrooms) are spaces open to animal companions; and this has been seen as a key characteristic of posthumanist households (Power 2008). Charles's (2016, 7) work indicates an absence of physical boundaries and the sharing of domestic space, including beds. Certainly, space is important to the canine inhabitants of a household. Kenneth Shapiro has suggested that dogs are spatially identified and grounded beings, and that the self of a dog is constituted through their association with space (Shapiro 1990, 189). Jeffrey Masson (1998) notes that dogs have a tendency to be in the present. Given both this and the spatial preoccupation of dogs, the impact of animal companions living in human homes in a humancentered world means both that conflict and accommodation is likely.

In the case of some species, such accommodation may be profound. For example, in her fascinating discussion of living with house rabbits in what she calls posthumanist households, Julie Ann Smith (2003) shows how members of House Rabbit Societies radically altered the space of home, or,

> 'rabbit proofed' their houses, a playful word that euphemized extensive modifications. In my own house, rabbit-proofing meant that most of the furniture was made of metal, electrical cords were fastened behind furniture or covered in hard plastic or metal tubing, and protective wood strips were tacked onto wood baseboards and wood trim around closets and windows . . . linoleum replaced carpet or the carpet was abandoned to shredding. (2003, 187)

Smith writes compellingly about the ways in which she and other keepers of house rabbits attempt to accommodate peculiarly rabbit concepts of the management of household space. While such 'extensive modifications' may seem shocking, modified space is an inevitable accompaniment of having animal companions in the space of home. Smith's proposal is radical—that a rabbit in the house is recognised as a full household member as opposed to a lesser being whose life will be largely determined by human needs and wants. Long-term living with large numbers of rabbits taught Smith, for example, that rabbits like to move along corridors at the edges of rooms, and thus rabbits will attempt to move all unfixed furniture to the centre of a room to accommodate this. This understanding, she says, 'changed forever the way I live in my house' (2003, 188). Things may be less dramatic in sharing space with dogs, but as my ethnography indicates, modifications there are, and also resistance to the assertion of dog understandings of the appropriate use of space and occasional conflicts resulting in battles of will, as we will soon see.

Despite this, animal agency in the home is highly qualified. The household is a site in which anthroparchal relations intersect with other systemic relations, of patriarchy, capitalism and so on. As commodities, dogs can be passed from one household to another, and possibly abandoned; such as Kevin, my research assistant, who came into my hands (literally) by being passed over the fence by a neighbour who was talking of giving him up to a shelter. The Western household is a key site of consumption, where the bodies of domesticated animals not considered 'pets' may be variously stored, prepared and consumed as food by both human occupants and companion cats, dogs and others. The tensions and ambiguities surrounding the eating of other animals in homes shared by dogs and humans will be the subject matter of chapter 4. As we saw at the start of this book, for many nonhuman creatures, a human home is a site of domination as cats, dogs and other animal companions may be implicated in forms of gendered domestic violence (Taylor and Fraser 2019). Alternatively, 'pets' may be neglected, or treated with cruelty or violence in this privatised space, relatively immune from public view (Fraser and Taylor 2021). Owners, along with their animal companions may also face discriminatory and exclusionary practices in relation to making home. For those reliant on private rental markets, prejudice and competitive renting conditions make it difficult for multispecies households to rent (Power 2017); while for those seeking to leave domestic violence situations, it is often difficult to find either temporary or permanent housing for human victims and their animal companions (Laing 2021a; 2021b).

Despite such restrictions, exclusions and forms of domination, in a human-dominant public world, the relatively privatised site of home might also be a site of resistance where the eating of nonhuman animals is rejected; where relations of care and conviviality might enable the relative flourishing of humans and animal companions and where practices are negotiated and human boundaries challenged. Relational ties between humans and animal companions are established though everyday interaction in home space. Companion animals share everyday human routines, and routines are adapted to incorporate dog preferences (Power 2008, 549) resulting in a 'lived intersubjectivity' (Fox 2006). In addition, the freedom of nonhuman animals in home space may be less restricted than in public space where dogs are subject to increasing forms of surveillance and control (Fox 2017). Dogs also adapt the space of home. Philo and Wilbert (2000, 13) suggest that there is a distinction between human constructions of 'animal space' and animals' constructions of space, which they refer to as 'beastly places'. These reflect animal understandings of the use of space and may be transgressive. The dogs in my ethnography certainly attempted to etch their 'beastly' marks upon homespace.

It is in this sense of the home as multispecies, as *also* queering species boundaries *and* as enabling elements of anthroparchal resistance or alterity, that I understand the homes of the subjects in this study to be posthumanist. As Ingold (2012, 173–74), points out, all creatures 'dwell', and while making home is a part of what makes us human, this may not be exclusively human[6] for the 'house also has many and diverse animal inhabitants—more perhaps, than we are inclined to recognize' (2012, 187). Yet it is not enough to describe home as multispecies. My use of posthuman is not to signal a new social form but to recognise social life *as ever* multispecies (Cudworth 2011, 8–10). In addition, it is important to recognise that animals are not simply to be added in or included, rather, the quality of home and relations within it alters in its more-than-human constitution. The space of home is at the same time dynamic while also routinised (Massey 1994, 168). This is where this chapter locates 'home'—in the intersection of power and its negotiation and of routine and shifting practices.

MAKING SPACE IN POSTHUMANIST HOME

In the processes of everyday routines and homemaking practices, the human-animal boundary is muddied, and even those humans who make stronger attempts to inscribe species hierarchy in the home breach it to some extent. The posthumanist home is muddy, but to varied degrees, as the maintenance of species boundaries is fluid and partial. The data analysis suggests that heterotopia is found in the everyday practices of multispecies living and in posthumanist homemaking, as we will now see.

Spatial Negotiations—Who Goes Where?

For the vast majority of interviewees, dogs were allowed free reign of the entire home, thus statements such as these were common:

> She follows me everywhere so she goes when I go to the loo. She sits outside the toilet. (F, London)

> He's allowed all over the home. (F, Leicestershire)

> Everywhere, everywhere, everywhere—yeah, everywhere—they've got the whole house to themselves. (F, London)

For the majority, few attempts were made to restrict where dogs might go. For some, this was a point of principle, enabling the free expression of dogs within the home, as one put it: 'I wouldn't want a dog that was so regimented that it can only sit in a certain place' (F, London). In some households, dogs had their own (human) sofa (couch): 'they have a sofa each, so they don't have to share and can spread out' (F, London). In a third of households, however, dogs were kept out of certain areas (particularly bedrooms, or in a number of cases 'the upstairs') or were trained not to sit on the furniture. Here, the spatial arrangements of home were subject to the drawing and often periodic redrawing of boundaries for both humans and dogs. Sometimes this is for practical reasons, such as the size of the dog or the number of dogs:

> Clearly, he's quite a big dog [the dog is huge, a mixed mastiff/Great Dane breed], so he's not allowed up on the sofa because then there'd be no room for anybody else, we kind of knew that. We sleep in the loft and he can't manage the stairs. . . . He's not allowed in the dining room when we eat and there are

> kind of doors that we can shut, exclude him from bits we don't want him to go, but he's pretty much got the, er, he's got the run of most of the downstairs of the house. (M, London)

> having so many dogs at once, I couldn't afford to let three or four or more of them just take over the house. (F, Leicestershire)

For others, the literal messiness becomes too difficult to handle:

> I will let the dogs on the couch although, you know, I don't let them on the beds anymore. I used to let them on the beds, but I've stopped that now because it did take over and, um, feet marks and hair everywhere and, you know, it's quite, well you know, she'll [the larger dog] sit on yer bed and have a couple of farts and she'll tread in poo and walk all over your bed [laughing]. (F, London)

For others, marking space for humans only is an issue of training and minimising 'unwanted' dog behaviour in the home, referring to their own knowledge and research into dog training:

> All the dog training books said that you need a separate space, that it's not good for them to sleep, um, in the bedroom or, erm, not be confined—they shouldn't have the whole space, so he now has slept in the kitchen, erm, ever since he was, say, six months old. Prior to that, erm, he slept with us in the bed initially and then we realised, reading the books, hearing from other people actually that was not the best thing you can do to, erm, build this relationship where you're supposed to be, erm, higher in the dog hierarchy, that he must not sleep in the bed so then we had to try to get him off the bed which was really difficult. (F, London)

> I have read lots of books that says that the top dog has to be the highest, higher, so if you sit here [on a chair], they would want to sit there [on the top of the back of the chair], they've got to be looking down on you, and so on. So up in the bedroom, if you give way, that then becomes their territory. We've never encouraged dogs upstairs or onto furniture. (F, Leicestershire)

> I don't have dogs sitting on laps—I think you have got to make a division between what is human and what is dog, and the dogs don't go on sofas and they don't lie on beds and things like that, and, you know, the furniture is not there for the dogs. (M, London)

What is important to note, however, is that dogs—as agential members of human households—attempt to circumvent human plans for their containment and management. Dogs are seen as constantly 'testing the boundaries' in homes where there are spatial boundaries. In some cases, dogs can be ingenious—in opening (and in some cases, also closing) doors or by escaping into gardens via cat flaps, open windows and so on. In this quote below, a man is engaged in an imaginary conversation with one of the dogs he lives with and with whom he claims he has 'a battle of wills'. The other dog he lives with is far more compliant, yet he sees that dog as the favourite of his 'wife and the kids' rather than 'his' dog and a worthy opponent in the struggle over domestic space. In telling the story below, it should also be noted that he cannot stop laughing:

> Now since the [child stair] gate's been taken away—because I actually found that I could put it in the garden then it could stop Lola [name of dog] going up the bloody wall and into next door's garden—so immediately that was taken from the stairs she said, 'Right, ok, I'm gonna go upstairs now' and . . . then she took to sleeping on the beds upstairs. So, the *barricade* I have to, have to go through if I come downstairs in the night [laughing], the *business* before I go to bed at night [laughing] it's a sort of, it's a sort of a battle of wills you can have with dogs, well I have [a battle] with this one anyway. So, I thought, 'Right, I'll put the ironing board on the sofa to stop her lying on the sofa'. I have to put [name of his partner]'s sewing basket on the chair to stop her sitting on the chair, then I have to get a folding chair and jam it between the stair rods on the stairs to stop her climbing up and lying on someone's bed, and I think, 'OK *now* try and get in' [laughing hard]. I suppose I'm about as childish as the dog, really. (M, London)

There was a gendering to human control of homespace. A number of heterosexual and cohabiting female interviewees did indicate they were more open to allowing dogs free reign in the home than their partners:

> I would allow her in every room, but my partner doesn't want her in the bedroom. She loves coming in our bedroom, and if we leave the bedroom door open when we're out, then she'll sleep on the bed, erm, and it's like she knows that's the holy of holies, you know, the sanctuary. (F, London)

In all cases, dogs were allocated clear spaces which were for their exclusive use within the home. All the dogs had some kind of official bed, sometimes in a dog crate (although none of the interviewees crated dogs beyond the 'puppy stage', many dogs seemed to like to have an open crate in the house), some had specific rugs or mats, chairs, sofas or even human beds which humans would not use. Most interviewees considered it important for a dog's welfare that they were able to retreat to a space which was their own. In this case below, a woman stresses the importance of the dog having an open crate in a spare room where he can escape from her partner:

> F: He couldn't bear to be separated from him [the dog], even for him to go to sleep, you know, so he had to [pause] you [directed at partner who is

shaking his head in disagreement] had to train yourself really to give him some space. You did.

M: Is this the same house we've been living in when you've had these dreams? Honestly, I [laughing]

F [interrupting]: He did, yes, he did! He was so overjoyed to have this dog, you know, umm, that he just couldn't bear to be separated from him, right from day one [laughing].

(Heterosexual couple, Leicestershire)

There is another space which dogs also occupy posthumously, the garden. While some interviewees buried dogs or scattered their bones (cremains) in public spaces such as a favourite park or field, many 'bring them home and they're all dotted here [buried] in the garden' (F, Leicestershire). Others kept dog cremains in caskets or other receptacles in the home.

The use of homespace, therefore, is open to dogs to differing degrees. Whatever the demarcation of space, the boundaries are likely to be flexible over time and subject to processes of negotiation. Negotiation is the right term here because whatever the relative power of humans in the home, the agential being of dogs asserts itself in the use of space. It is not always humans whose interests prevail.

Beastly Place: 'We've Put a Blanket over It Now'

In her moving and thoughtful discussion of the end of life of a much-loved dog, Jessica Pierce (2012, 56) suggests that caring for an aging dog places especially intense demands on those of us who wish to do as well as we can for them. We must adapt our homes and our routines very significantly to enable them to get by and put up with large amounts of mess of various kinds. Yet dogs, of course, also adapt spaces and after the death of her dog, Pierce reflects that he

> has etched himself into our house, through the stitched-up scars in the couches, the extra tall fences, the scratches in the doorframes, and the fact that every blanket and bedspread has holes. . . . And he has etched himself into my heart as fully and painfully as any creature ever has. (2012, 224)

Despite the large numbers of people sharing their domestic space with dog companions, there is very little if any discussion about the ways in which home space is marked by dogs. The destruction of furnishings seemed to be seen by the interviewees in my study as an occupational hazard of living with dogs for those who raised this as an issue. What is interesting in these stories of the destruction of the home is that tales of 'bad' behaviour are always mitigated. The ultimate mitigation is love, as the quote above from Pierce suggests. But I also found tales of canine improvement or human adaptation:

> Up until the age of about two, he chewed through two settees, a kitchen wall, he rounded off all my kitchen cupboards [cupboard doors] but, um, he grew out of that . . . he's the perfect dog now. (F, London)

> We had new carpets maybe a year ago. Of course, then he was a puppy, and Max squeezed a sample tube of foundation that I got off a magazine or something, and he ate it in the lounge, squeezed it all over the carpet. Fortunately, the insurance said it wasn't the dog's fault and so they're replacing [the carpet]. (F, Leicestershire)

Alternatively, mitigation takes the form of human resignation to living in a home with damaged furniture:

> Our couch, well, it's been repaired once, but it's completely wrecked. You know, structurally its fine but you know, the cushions and seat and stuff—we've put a blanket over it now. . . . I like tassels on cushions, but I can't have those now, you know, in case he swallowed them. And I like cushions, expensive cushions, and the three we've got left are all, you know, mangled. I come back from work and all the corners are damp and look weird. And with the first cushion that was trashed, I got annoyed, but now, well [pause] you know. (F, London)

> Max used to strip the wallpaper off and then Freddie started doing that. And he's eaten nearly the whole of my big window seat . . . pulled all the stuffing out so there's a big hole. My husband said, 'You'll have to renew it'. I said, 'No, not until he's grown up', so I keep putting a blanket over it now. (F, Leicestershire)

There is, thus, a literal covering up of the household damage caused by dogs, and a metaphorical covering up whereby this is described as behaviour which is 'grown out of' or was/is expected to be temporary.

Another way in which dogs assert themselves in marking homespace is by considering what is supposed to be for humans to be something for them. In most cases, this involves food, but occasionally, clothing and particularly that which is often left on the floor and found by dogs—footwear:

> She ripped through some lovely boots from [name of an expensive shoe shop], red, leather, ooh I loved them, toes straight off [laughs]. Kept them for ages and ever-so-often I'd get them out the cupboard and go [to the dog], 'Look at the boots!' [laughs]. It wasn't even just chewed a bit so you could still wear them, it was chewed off. (F, Leicestershire)

> She chewed my friend's shoe to bits, and it wasn't a cheap shoe. The trouble was that I didn't catch her doing it. I only found the remains so I couldn't really tell her off because she'd gone by then, she wouldn't have known why I was telling her off or whatever. So yes, that cost me quite a bit of money, to replace the shoes. (F, Leicestershire)

Stories of the destruction of clothing are usually laughed away, and often caveats about clothing being left 'out' or 'on the floor' serve as excuses for destructive behaviour. These stories of destruction are always found to be (retrospectively) amusing with people keen to tell stories about them, and the most laughter surrounds tales of dogs stealing food. Here, interestingly, no caveats or mitigation are added because food is assumed not to be left in reach of dogs, or dogs are assumed to understand that worktops in kitchens, the tops of ovens and tables are spaces from which they are excluded.

> Well, the Staffie [Staffordshire bull terrier rescued] from out of the canal, he could open the fridge. And he'd do it while you were there. He got so, you know, just like he'd be standing there and [impersonating dog] 'move along', pushing you out the way. I used to do quite a bit of catering and she [another dog] had off the side [kitchen worktop] seventy-two vol-au-vents, seventy-two [laughs]! Luckily, they weren't filled. (F, Leicestershire)

> He stole twelve onion bhaji's and twelve samosa's—frozen—from on top of the cooker. I was laying the table, for a meal, and I heard this big clang—and it was on the back of the cooker as well—a big clang on the floor. By the time I got in there, nothing left, he'd eaten the whole lot. (F, Leicestershire)

> I'd done a huge tray of sausage rolls, puff pastry, pork sausage, and they'd all gone. I turned the tray round so the hot ones were at the back, and I shoved them right at the back [of the worktop]. When I come in, well I bet she'd ate, what? three dozen? She was scuttling about on the floor, 'It weren't me, it weren't me', trying to eat the crumbs up. (F, Leicestershire)

> The other week—done it a few times—we've left the French bread out for the morning, forgetting we've got a stealer, and we had a lay-in—Aisla doesn't like you having a lay-in. In the morning we forgot there was a baguette on the table because it wasn't there anymore. The packet was on the floor. (M, Leicestershire)

Dogs also have their own sense of what makes a pleasantly smelling home and a desirably perfumed body that are often starkly at odds with human notions. Here, human disgust at dog behaviour is clearly apparent and strongly expressed:

> Rolling in fox's poo, eating it, stinking out the car. When he trumps at home, we all blame each other, but it's 'im, lying on the floor [laughing]. I was always told that dogs don't fart out loud, but we've 'eard 'im, and it's just so funny innit? Childish, but funny. *And* they look at the person nearest them as well. (F, London)

> Well, he farts a lot, pretty much constantly when he's dosing, which if he is not out walking or eating, is pretty much most of the time. I mean he just creates an atmosphere [laughing] a thick fug, a thick fug of horror [pause for laughter]. And always nearest to me as well. And sometimes, well bloody hell, phew! God. A thick, thick fug. (F, Leicestershire)

> Billie does have the habit of eating rather disgusting things on occasions and feeling somewhat the worse for it and on other occasions actually sort of deciding to retch them up in the middle of the living room which is just so *vile*. I've actually . . . put rubber gloves on and brushed her bloody teeth cos, you know, rank bloody breath and everything, you know, 'You're staying there! Stay out there!' [directed at dog, laughing]. (M, London)

> It's constant, constant. You take your eye off of him for one fucking minute and there he is—covered in Channel No. 5 [fox shit]. He loves it. He just *loves* it. I wash him, of course, tomato ketchup,[7] shampoo, in the bath or under hose, but it just lingers. I think he pretty much always smells slightly of shit. My house must stink. (F, Leicestershire)

> I found he wasn't very good with being house-trained with his wee, it has to be said. That was rather a consideration that drove me to despair, but he is OK now, obviously . . . at the time I just remember weeping and constantly mopping and just thinking, 'Oh my god, oh my god, this is awful' [laughing]. (F, London)

Damage to the house and 'disgusting habits' such as rolling in or eating faecal matter were often seen to be inconvenient and a hassle to deal with, time consuming or expensive, but very few people considered this to be the most negative aspect of home lives shared with dog companions. For those with children or grandchildren, some of the more 'disgusting' aspects of embodied living with other animals were compared to looking after children, with many comparing 'picking up the dog poo' to changing a nappy/diaper:

> I suppose I've been kind of bringing up kids for so long that in some ways it's like—you know—it didn't feel like a big deal to take on a responsibility of a dog I've sort of been bringing up children for the last thirty years, more or less. I've got sort of step-sons as well as grandchildren and things . . . mopping up after other people's mess is, well, I've done it for so long it doesn't really matter whether it's a human or a, or an animal. (M, London)

> I mean, a lot of people say 'I'm not having a dog because I've got to pick the poo up', but I've had two children, I've done nappies, and you know, you can wash your hands. But I have to pick the poo up, he [her husband] won't do it [laughing]. (F, Leicestershire)

There was a gender difference when it came to clearing up mess in the data from the rural sample, as the quotes above, both from heterosexual people of a similar generation, exemplify. Amongst the London interviewees, both men and women regularly cleaned up after dogs.

Rather than destructive behaviour or dealing with unpleasant smells, vomit and shit, it was the impact of human working lives on dependent dogs left in the home that was often referred to as the 'worst thing' about living with a dog, and this will be a focus of the final chapter of this book. It is interesting to note that other studies have found that one of the main reasons people have given as to why they do not have a dog companion is that they are working or out of the house most of day and/or do not have enough time for a dog, rather than not 'liking' dogs (Westgarth et al. 2007, 2). Most of the interviewees in my study were very conscious about the length of time dogs may be left alone, and many were concerned dogs could become 'bored' or 'lonely' while home alone. A number of interviewees commented that any destructive behaviour (of furnishings, for example) may be linked to boredom. As we will see in chapter 6, the most popular answer to the question 'What would make your life easier living with a dog?' was a version of 'not having to go to work'. The second most popular answer involved the climate in the UK, for example, 'Not living somewhere where it rains all the time', and this has direct bearing on the issue of living with mud.

A Less Clean Space

The majority of interviewees in this study were women, and only one of the men interviewed mentioned the presence of hair, dirt and mud on the home, or its impact on reproductive labour. Most of the women, whether single or cohabiting with other humans, raised the cleanliness of the house as an issue. However, there were distinctive ways in which the 'mess' dogs bring in and the way it is regarded are discussed.

For some, there was a story of rebellion and of a positive rejection of being 'house proud', with some suggesting that 'if you're particularly house proud, don't have a dog' (F, Leicestershire). This was often related to discussion of relationships with parents (mothers particularly) and, in some cases, a desire to be different from them:

When we go [to visit parents] with the dogs, Mum tries her hardest not to care about the hair and stuff. . . . She puts down these sheets and they sit, but not quite on the sheet. And she says, 'Why don't they sit on that sheet?'. And I say, 'Well they're dogs, Mum, you can't expect that'. So, she's kind of given up on that idea now. We're not house proud at all. (F, London)

My Mum didn't like cats and she didn't want a dog anymore . . . I think it was the hair and the dirt—very house proud. I think that is why I like having lots of animals and hair everywhere, it's a little bit of my rebellion that's still coming out [laughing] against order and tidiness. (F, London)

I always wanted a dog, but my Mum and Dad would never let me have a dog because they were very house proud and my Mum and Dad have always done whatever they've wanted and I think, in this day and age, they wouldn't have had children, you know, very selfish. We always had cats and we lived in the countryside, but we could never have a dog, never *ever* have a dog. So, as soon as we got married, the first thing I had was a dog, and then I had two cats. (F, Leicestershire)

my Dad is an animal lover. My Mum would never have had any pets, and she sees all the practical problems like the extra housework and going on holiday and stuff. (F, London)

For a small number (three cases) of those exclusively in the Leicestershire sample, designer crossbreeds with poodle genes (labradoodles, goldendoodles, cockapoos) or other breeds that did not shed hair (Tibetan terriers) were seen as desirable dogs with whom to share home space because of their lesser impact on domestic labour:

M: There were certain conditions about having the dog, like it couldn't shed [laughs] because that was the fear of having hair all over the place wasn't it [directed to partner]?

F: Well I'd [pause] umm, yes, because our son-in-law has got, umm, asthma and certain allergies and then the grandchildren as well, and also . . . we'd had a young woman come to visit us, who was blind, with her guide dog, a Labrador, and the Labrador didn't move from her feet for an hour and when she left the floor was covered in these dog hairs, and I said to [name of partner], 'I'm not sure I could cope with that'.

(Heterosexual couple, Leicestershire)

I was interested in the Tibetan terrier because another friend of mine had one and they don't moult and they're very good for people with allergies and I've

> always suffered from eczema and my sister's allergic to dogs, but not to Maja [her dog]. (F, Leicestershire)

Here, the desire for a dog which does not contribute hair to the house is run alongside comments about allergies, perhaps to legitimate concerns about a hairier home. For others, living with a dog leads to a shift in their ideas about the cleanliness of home space

> I suppose I was a little bit house proud before I had him, a bit particular, but that's gone . . . you just accept that you're going to have muddy footprints and, you know, you're going to have bracken brought in and you're going to have muddy paw prints within a day of cleaning your windows, you know, but it doesn't matter, I mean, it *doesn't matter*. I'd prefer to have Rio than a clean and tidy house all the time [laughs]. (F, Leicestershire)

For the majority of interviewees, mud and various other forms of 'dirt' were inevitably associated with living with dogs. Most female interviewees mentioned paw prints, damp dog smells, splash marks on walls from wet dogs shaking themselves in the house and that their homes were sometimes a bit more 'smelly' than they would be without a dog in the home. Such observations were accompanied by a range of accommodating comments such as 'that's what they do and that doesn't really bother me' or that people simply have to make extra time for cleaning the dog(s), themselves, clothing, towels and household space after wet walks, which in the UK is considerably often. In most cases, detailed stories of the management of dirt within the home are often followed by matter-of-fact statements about the inevitability and acceptance of muddier home life:

> I've got a boat and a cottage. At the boat, there's not a lot I can do so she just goes in, and you know, poof, she's everywhere! At the cottage, I've got more of a chance, so if she comes back and she's really, really muddy and wet, I could put her in the bath and give her a shower which she actually quite likes. It's very rare that I do do that, but I can do. And I do have a blanket on the sofa so she can go on there, but she doesn't have to. If she goes on where there's not a blanket, I don't tell her off. But you know, that's just one of those things that I came to terms with as soon as I got a dog. You know, I don't want to have a dog and then say, 'Oh you can't go here' and 'You can't go there'; I've got a dog and I just deal with it. (F, London)

For most of the interviewees in this study, therefore, while the extra work dealing with the stuff brought into the house by dogs is always seen as something that people could 'do without', it is also seen as something to be taken on when people decide to live with dog companions. A concern for a number of interviewees in both study sites was that people adopting dogs perhaps do not always realise the amount of work and time this is likely to involve, 'how life changing it can be. And they give up too easily, you know, it is difficult' (F, Leicestershire). Finally, many households attempted to accommodate the messiness of living with dogs by adapting their homes—for example, some removed carpet and replaced this with tiled, wood or laminate floors; others elected to have furnishings with washable covers. Dog-proofing may be less demanding than the rabbit-proofing discussed earlier, but nevertheless required quite significant adaptation to accommodate 'beasts' in homespace.

'Just Put Another Dog on the Bed!'—The Dirty Pleasures of Co-sleeping

A particularly peculiar form of mud was seen as that which accrued on and in the beds that humans shared with dogs. A number of interviewees pondered the strange nature of mud wondering whether it is 'basically sand' or 'some form of ground glass—that's what it feels like when I get in bed!'. Despite the literal muddying of sleeping space, however, a significant proportion of interviewees shared bedrooms or beds at least some of the time. 'Sleeping with your Jack Russell [terrier], what's not to love?' (F, London) asked one interviewee, slightly indignantly. In most cases, people did seem to enjoy co-sleeping; dogs were usually allowed to sleep in the bedrooms and more than one third of my interviewees shared their bed with dogs.

Interviewees described this sharing of sleeping space as 'a free for all', or described where in their bed a dog might habitually sleep: at the bottom of the bed, across the pillow, on one side of the bed, between a human couple or surrounding a human: 'one sleeps at one side, one sleeps the other side'. This pattern of co-sleeping was maintained even if other members of the household were not so enthusiastic, or when space might be in seriously short supply:

> Billy sleeps on the bed and the cat sleeps on the bed as well. Sometimes [her son] gets in as well. Sometimes it's just me and the dog and the cat and my husband gets fed up and goes and sleeps in the other room. (F, London)

> the boys [dogs, all lurchers] all sleep on the bed [Erika—'All 3 of them?']. Yep.

> [Partner interjecting]yep, the three boys, always on the bed, fighting to get in. Meg sleeps downstairs in her own basket. (Lesbian couple, London)

All co-sleepers had stories—that will be familiar to anyone who has attempted to share bed space with dogs—of a dog's ability to acquire and maintain space, and to move considerably larger and heavier sleeping human bodies:

> I've got a king size bed, and he takes up most of it, so I usually find that I'm across the top or wedged into a really small bit of the bed and he's stretched out over the whole thing, and he takes up a surprising amount of space for a small dog [meaning slim, a whippet lurcher]. But it really amazes me that he's so heavy as well so I can't really move him either; I have to actually kind of wake him up and get him off the bed, erm, just so that I can move. (F, London)

For most co-sleepers, there is some sense of guilt attached to sleeping with dogs, perhaps deriving from reading books on dog training (which almost invariably advise strongly against this). A number of interviewees describe co-sleeping as a form of guilty pleasure and certainly some were aware that other dog owners and certainly those not living with dogs would find co-sleeping difficult to understand or perhaps would be revolted by it as inappropriate or unhygienic:

> He mostly sleeps with me which I always said that I wasn't going to do but then it's just too tempting, you know, he comes in for a cuddle in the morning or something. It's really nice. I have a cup of tea and a cuddle on a Saturday morning and take him out for his walk. (F, London)

> People say, 'Ugh, that's disgusting', you know, but I think, 'What *are* they going on about?', you know [laughing]. I would rather have some dogs in my bed than some of my friends, you know, with the mess they make, and you know, the carnage they cause! (F, London)

> M: When you speak to people who have got dogs and they go, 'What? You let them sleep on your bed?', they look at you in such a disgusted way.
>
> F: No, but it's not like we make him, it's like you know
>
> M [interrupting]: Obviously you don't make him, he chooses to, it's that thing of being close, it's like pack animals would sleep together, it's that thing. He sleeps between us, on his back, snoring his head off.
>
> F: Occasionally, on a very warm day, he might sleep on the floor near the window, but no, he likes to be with us, doesn't he?
>
> M: Yes.
>
> F: He likes to be close.
>
> M: Yes, let's just say [pause] he's a bit of a mood-ruiner.
> (Heterosexual couple, Leicestershire)

The pleasure of a 'cup of tea/coffee and a cuddle' in bed was a feature repeated across almost one-third of interviews in both study sites—even people who did not regularly co-sleep at night, would allow a dog on the bed for such a morning ritual. Others described the pleasure of a warm dog in a bed, one with a story of staying in a caravan in winter and being warned by the warden of the site that it might be too cold to stay—'Don't worry', she quipped, 'We'll just put another dog on the bed!'.

Some people were highly resistant to sharing a bedroom with a dog, let alone a bed, but another common tale is one of a dog's ability to be persistent. A number of interviewees who regularly shared their bed with dogs commented that they had not intended to do so at the start of their relationship. Interestingly, for a number of those co-sleeping, the notion that dog sociability needs to be taken into account comes through strongly—as the male interviewee in the above quote suggests, in the posthuman home, the happy pack is the pack that sleeps together. Here below, a woman has specifically been given 'expert' advice on pack co-sleeping:

> He barked and howled for a year every night when I put him to bed, and he would go all night without stopping. He did not stop unless I went downstairs and sat with him. So, I used to go downstairs because I couldn't stand it in the end, and I was worried about the neighbours as well. So, I used to sit, write my essays, do my studying, slumped over the little kitchen table that I had, and I wouldn't look at him, wouldn't touch him, wouldn't fuss him, and I did that for a year. And in the end, I just thought, 'I can't do this anymore'. And I'd spoken to, umm, a lady in Germany on the Schnauzer website and she said, 'The worst thing you could have done was try to make him sleep on his own'; she said, 'They like company, they like being with you, they feel it's their job'; she said, 'You should have just let him have a bed next to you, and you wouldn't have had this'. So one night, three o'clock, I just . . . said, 'Right, get upstairs then', and he just looked at me, I looked at him and he was just so shocked, so I let him go upstairs and that was it. (F, Leicestershire)

Finally, it is worth noting that within the London field site where a number of lesbian couples and quite a few single women were interviewed, these women were likely to allow dogs in bedrooms and usually also to sleep in their bed. Amongst heterosexual interviewees, there was illicit co-sleeping by women if their partner was away. This did not characterise the data from the Leicestershire sample, however, where the key factor that seemed to influence the likelihood of co-sleeping was the number of dogs and/or the size of dogs. The tolerance of people to sharing bedroom or bed space with dog companions, was an unanticipated finding.

Home Is Where the Dog Is

> And you never come home to an empty house. And they greet you, and they're pleased to see you, no matter what's happened, they love you. I can't imagine being without them. (F, Leicestershire)

A number of interviewees discussed concerns about having dogs in the home. For some, this involved anxieties, past or present, with renting properties and wanting to have or having dogs. For some, where they were able to live was very much influenced by their decision to live with companion dogs. Many made the point that either a house, home or a family was not complete without a dog and both those with and without children expressed the opinion that it was a positive thing for children to be raised in households with dogs. Some emphasised that people need to be 'sensible', particularly with young children in households with dogs. A few commented that they felt 'safer' with a dog in the home, particularly if they were alone or if they lived alone.

All interviewees made some sort of comment about the quality of home and home life being altered in significant ways by the presence of a dog, such as that above. Many interviews contained a version of the quote above indicating that 'having a dog makes a house a bit more homely', or 'the animals make a house a home', and the homeliness was very closely linked to the presence of 'somebody who's always pleased to see you'. Most people commented on the way in which dogs greet them returning to the home and the constancy of presence of a dog in the house

> when you come home and there's somebody there that's pleased to see you. That's a huge thing, isn't it? Well it is to me anyway, you know, I come in and there they both [a dog and a cat] are, you know, he's [the dog] running round and he goes to fetch a toy, and he's so pleased and so excited to see you that even if you've had the most rubbish day that you can imagine you just think, 'Ah, it's great to be home'. (F, Leicestershire)

Alternatively, a house without a dog is seen as a more 'empty' or 'lonely' space: 'The house would be very empty without a dog I think' (F, London). Some interviewees made the point that in the 'gaps' between dogs and after the death of a dog they had not wanted to return home after going to work, on holiday or out of the house even for a short time such as for shopping. In one case, a couple described a period of a year without a dog as one in which 'we ate out, stayed out late and kept going away just because we really just didn't want to go home. We couldn't be in the house' (Lesbian couple, London).

All interviewees talked about sharing space with dogs in the home and engaging in shared practices—for example, some described specific games they had developed in the home with their dog:

> We play this thing, well, we call it stair-ball because she sits at the top of the stairs and tosses her ball down or her toy, whatever it is, and barks until you go back and throw it upstairs. (F, London)

Others watched television with their dogs, or their dogs would be with them in ways which interviewees saw as engagement, be it watching them cook or undertaking housework or repairs, 'helping' with gardening or, in one case, meditating. Some interviewees referred to a dog's feelings about their 'family' or 'pack', emphasising that dogs seemed to be more relaxed or 'happy' when all the members of the household were together. Many claimed that the presence of dogs was 'relaxing' or 'chilling', and perhaps this ability of dogs to live in the present and enjoy the moment is part of what enables them to exert a calming influence.

These aspects of sharing home are important cues not only in ascertaining the close relations people have with dog companions and the interactions they have but also the discrepancies of power. The idea that the dog at home waits for humans—to provide attention, affection, company and entertainment—suggests that the overwhelming beneficiaries of this relationship are human owners. However, the material described here paints a partial picture. As we saw above, dog owners are highly concerned about the impact of their working lives on the life of the dog who lives with them. As we will see in chapter 6, those living with dogs considerably adapt working times and patterns and leisure pursuits (beyond dog walking) to ensure a fuller and more varied life for the dogs of their hearts. Thus, while it is difficult to conceptualise ownership of dogs and their keeping as companions as anything else but an exercise of dominatory power, the humans in this study were aware of the unfree condition of dogs and concerned to make home in ways that reflected their understandings of dog interests.

A MONGREL DOMESTIC

The difficulties of sharing living with dog companions involve increased levels of cleaning and washing, drains on time and, for many, a deleterious impact on the home in terms of damaged property and the creation of mess. My human interviewees were clear that lives with dogs involved time, work and effort, and that, at times, these were constraints one might 'do without'. In these narratives of daily life, there is negotiation and renegotiation about the use of space and the boundaries of what is, and is not, human and what is, and is not, for dogs, and these boundaries are very often porous and dynamic. Many of the problems faced by those sharing lives with companion dogs are seen as somehow inevitable or are mitigated in some way—often through physical adaptation to the home or adaptations in human concerns—which makes dog being and dog behaviour acceptable. An emerging question, and one which is a focus for chapter 6 is the extent to which this nonhuman animal-related work can be understood in relation to feminist conceptions of domestic or reproductive labour. There were certainly gender differences in terms of a concern for the work involved in caring for dogs and in sharing home space with them. Despite this, women were more likely to allow dogs free reign of domestic space and to share (or wish to share) intimate space with them.

While some homes allowed individual dogs free movement, in others there was some policing of the human distinction through boundary marking in the space of home. Yet dogs, as active shapers and shifters of life in the home, assert their presence across the domestic space: stealing food from kitchens and inches from beds; asserting their wills in the use of beds and bedrooms, chairs and couches. Many dogs attempted to make homespaces beastly places, incurring increased levels of cleaning, washing, drains on time, and damaged property. The posthumanist home for my human participants is a space in which dogs contest boundaries, both physical and emotional and where everyday practices and relations are muddied. In the breaching and reinscribing of the boundaries of species and of domestic order, there is lived intersubjectivity, played out in multispecies routine and homemaking practices, intraspecies communication and the cementing of affective ties.

What has been little discussed in this chapter is the unfree, dependent and vulnerable position of dog companions in a social world that is human-centred and human dominant. Yet within the narratives of the 'responsible', engaged and concerned dog owners I spoke with, power makes its presence felt in more subtle ways, for example, in the uneasiness many feel about the status and lives of animal companions often left in the home for long hours by humans leading exclusively human lives at work, in the supermarket or at dog-free leisure events. It is perhaps this unease which prompts the array of caveats and mitigation which cover the difficulties of posthuman home life. The posthumanist home is not a space of equality but of human domination, and for some domestic animals captive in the home, it is a place of deprivation or violence. Yet the dogs in this study were, as Varner (2002) suggests, mostly domestic partners, otherwise animal companions and never 'mere pets'. There are degrees of intimacy present, but every interviewee had something to say of how lives, households and the experience of 'home' is altered by the presence of dogs and how their relationships were with particular dogs as individuals.

There is not purity of species in the posthumanist home but what might be described as a 'mongrel domestic' (Cudworth and Jensen 2016) in which human(s) and dog(s) come together in various degrees of spatial, physical and emotional intimacy. In muddying dwelling spaces and domestic relationships, dog companions are active in making a different kind of home. Despite anthroparchy in the public world and abuse and neglect in many homes wherein dogs are kept, the posthumanist home can be, at least in some cases and to some degree, a 'haven in a heartless world'. For both human and canine dwellers, home *can* be a place where love and care happen, and dogs can perhaps express selfhood more authentically as a lively presence attempting to shape their place. I found what I consider to be glimpses of heterotopia in the spaces of muddied living, something to which we will return at the close of the book. In muddying our dwelling spaces and our relationships, dog companions are active in making a different kind of home—a mongrel domestic place.

Chapter 4

Eating Animals

Dog companions kept in homes are overwhelmingly dependent on humans providing their food. Eating is a vital activity undertaken daily in the household. As we have seen in previous chapters, people may walk together, sleep together and otherwise share homespace with dog companions; however, they do not eat together and rarely share the same food. Rather, what dogs eat is commonly referred to as dog food, and in the UK, as elsewhere, feeding animal companions is big business. The 2023 Annual Report of UK Pet Food (previously the Pet Food Manufacturers Association, PFMA) found that people from 57 per cent of households shared their home with an estimated 12 million dogs and 11 million cats (UK Pet Food 2023, 1). Providing food for these animals draws in significant revenue, with the UK pet food market estimated to be worth £3.8 billion. Considerable expenditure is outlaid in feeding dogs, cats, small mammals and fishes, kept both outdoors and in, but of these markets, dog and cat food products dominate by a considerable margin. The income generated by dog food and related products was estimated to be £1.84 billion in 2023 (a rise from 1.60 billion the previous year) and from cat food and related products was £1.43 billion (a rise from 1.27 billion the previous year) (UK Pet Food 2023, 1). The development of specific food for animal companions, however, is a relatively recent development.

Until the late nineteenth century, domestic dogs were fed leftover food from human meals and stale or rotten food, and this is still common practice in countries where dogs are predominantly free roaming. From the development of dog 'biscuits' in the 1860s, a global industry has emerged which utilises enormous quantities of readily available corn, wheat, rice, potatoes and soy as well as huge amounts of blood, bone, skin and flesh from domesticate animals slaughtered for meat. In 2006, the pet food industry in the United States alone sold more than $16 billion worth of goods (Williams and de Mello 2007, 231). While there are various manufacturers of pet food, the industry is dominated by major food companies, and pet food has become a lucrative market for extracting profit by recycling gluts of grain and animal parts deemed unsuitable for human consumption. The UK pet food industry uses an estimated five hundred thousand tonnes of meat and poultry and five hundred thousand tonnes of cereal each year, using 12 per cent of the tonnage of by-products from the human food chain (UK Pet Food 2023, 1). Pet food, therefore, is an important element of the global political economy of meat, playing a

> critical role in the economic viability as well as the sustainability of the meat sector. Pet food adds value by taking the parts of the carcass that are not used (unpopular cuts such as the lungs, heart etc.). Without this, livestock production would be even more challenging. (UK Pet Food 2023, 2)

This link between the farming of some animals for food for both humans and the animal companions which they keep is brought into sharp relief by the logo fronting the 2021 PFMA annual report, depicting representatives of the five categories of animal that are the core consumers (a dog, cat, rabbit, bird and fish) facing core sources of their food (a cow, pig, chicken and plant). It begs the question, as all are encircled within the 'animal health and welfare' logo of the UK pet food industries, which animals and whose health and welfare are at stake when one is food for another? This said, the image also reflects the narrative of welfarism that has been extended from the health and welfare of the animal recipients of pet food to those other animals who become food for those kept as companions. The representation of those creatures which, as food for dogs (and for cats), are commonly represented on pet food packaging (as in figure 4.1).

Pet food is the subject of vigorous advertising encouraging human consumers to buy products that constitute a healthy, natural and/or economical diet for their companion cat or dog, and thereby demonstrate care and responsibility. Overwhelmingly, as we will see, this advertising naturalises food based on animals raised for meat, for cat and dog companions. Research on the relationship between ownership of animal companions, and concern with the treatment of nonhuman animals more widely, is inconclusive, although there is some evidence that living with a companion animal may encourage empathy to other kinds of animal (Serpell and Paul 1994). The ethnography generally found that many human participants did demonstrate concern beyond their animal, with the treatment of dogs more generally (dog fighting, the use of dogs in baiting and hunting wild animals, animal breeding) and animals in agriculture (particularly intensive farming). Although the majority of my interviewees were meat eaters, many did place some restriction on what they ate in terms of the creatures they would eat, or the methods of production involved in the animal derived foods they ate. While many were troubled by the 'what we love/what we eat' distinction, even the vegetarian

Figure 4.1 Dog food packaging with cow.

dog owners and the lone vegan in the study all fed their dogs meat. In this context, this chapter explores the role of carnism—the cultural naturalisation and normalisation of eating meat—in the feeding of companion animals. Carnism was operationalised unproblematically in discussing dogs' food and diet. Narratives of human choice infuse practices of dietary resistance to carnism, whereas biologism effectively reproduces the love/eat distinction when it comes to feeding animal companions. As one interviewee put it, 'I'm a vegetarian, but she's a dog'.

The first two sections set the context. First, there is a short mapping of the development of pet food and the changes in the way it is presented to owners as consumers, ending with a product recall which threw relations between humans, pets and those few species that are routinely eaten into sharp relief. Second, there is an examination of the normativity of dog companions consuming a limited range of animals in contemporary (Western) cultures through the concept and practices of 'carnism'. This, Melanie Joy (2010, 28–29) suggests, is so pervasive in our culture that we see it as routine, normative, apolitical—meat eating—whereas eating plants is constituted as a political choice for vegetarians and vegans. This section looks at the ways in which carnism shapes the narrative tropes of pet food advertising and paints the cultural backdrop which dog owners are influenced by, and negotiate, in making decisions about what is eaten, and by whom, in the home. The second part of this chapter uses material from the ethnographic research in order to argue that carnism pervades discourses of care and concern for much-loved dogs and that there are often also ambiguous relations to the animals such dogs eat. It will be argued that, even for vegetarian and vegan dog owners, carnist discourses reinscribe meat eating practices for the nonhuman members of the multispecies household. This chapter ends by considering the possibility of the vegetable eating dog, and the challenges for the future of the pet food industry.

DOMINANCE AND AFFECTION IN THE MAKING OF PET FOOD

As we have already seen, for Yi-Fu Tuan (1984), it is a complex relationship between domination and affection that leads to the historical development of different categories of pet. It might also be said, however, that similar forces are key to the development of pet food. There is a fundamental ambiguity at the heart of our treatment of the small range of species kept as domesticates in the West/Global North. On the one hand, animals are exploited in their billions for food and, on the other, a small minority of domesticates are kept as companions in homes where they may be unlucky enough to be mistreated or lucky enough to have owners who care for them greatly (see Cudworth 2011, 5–8).

There are ways in which the keeping of animals as companions both replicates, in minor form, the exploitative breeding conditions of the meat industry (Williams and de Mello 2007, 236–47). In addition, livestock farming and pet keeping are integrated forms of human domination of domesticates, for the breeding and selling of animals as pets ultimately supports a multi-billion dollar/pound pet food industry which is an important and highly profitable part of twenty-first-century agribusiness (Nibert 2002, 92–95). Certainly, every British slaughterhouse has a pet food bin or the 'McDonalds skip',[1] or for the meat deemed unfit for regular human consumption (Cudworth 2008). The amount of resources and the potential profit to be obtained from them is considerable as:

> Whole mammal bodies (cows, sheep, pigs etc.) by the hundreds of thousands, millions of their major parts and thousands of tons of bird flesh end up in the reject pile. Much of this waste is diseased. . . . This reject pile is soon on the move as it magically changes into 'meat meal' and 'by-products' on pet food labels. (Peden, quoted in Nibert 2002, 94).

Pet foods have always been produced from the leftover parts of animals which are not to be used for human food, such as bones, organs, ears, skin and so on. However, the industry expanded significantly from the mid–twentieth century. In the 1960s, most pet cats and dogs were fed dry biscuits, table scraps and some standardised forms of canned food produced from slaughterhouse by-products (Franklin 1999, 90). By the 1980s, pet food manufacturers were realising the potential of diversification and offering an array of brands, all produced by a small number of firms, to appeal to a range of human (rather than nonhuman animal) consumers (see Nestle 2008). In the twenty-first century, diversification proceeds apace with an increased array of luxury brands. In some cases, dog food has been endorsed by dog celebrities. After Eric Knight's novel *Lassie Come Home* was adapted into a feature film in 1943 by Metro-Goldwyn-Mayer (MGM), Lassie, the rough (long haired) collie star (and various offspring), became a dog celebrity, with sequel films, a radio show, a long-running train of television series running from the 1950s to the 1990s, and a remake of the original film in 2006. Lassie, both as an icon and a trademark, provided endorsements for pet foods and pet care, demonstrating global appeal by becoming the first 'canine ambassador' for the international charity Save the Children (*The Times* 2014). Since 2012, the famous collie is the brand property of DreamWorks Animation. Lassie is no longer a performer on the big screen, but is a merchandising star for dog food, accessories and training (Barnes 2014).

As we have seen above, pet food is a key element of the global meat industry and the consumption of pet food draws even those owners who do not include meat in their own diet, into the animal-industrial complex. The idea of an animal-industrial complex (A-IC) drew on C. Wright Mills's (1956; 1958) concept of the military industrial complex and was first developed by Barbara Noske to capture the extent to which human-animal relations are industrialised and 'embedded in the capitalist fabric' (1997 [1989], 22). The A-IC involves all the interrelated industries of agriculture (breeding and raising, feedstuff production), meat and dairy product manufacture (slaughter and processing) and distribution and sale through which animals become food. The business of animal research is part of the complex, as are financial interests, governing institutions and hi-tech, pharma and the military. Crucial to the complex is its political constitution. As Noske suggests, the systems of animal agriculture do not develop in response to public demand, but are part embedded in networks of industrial capitalism (from research, investment and financing to those of production and consumption) and produced by transnational corporations, such as Unilever (1997, 23). She describes the complex as 'ever expanding and self-perpetuating,' meeting needs that Western consumers learn to perceive as 'real' (1997, 38). Key to Noske's argument is that the mass exploitation of nonhuman animals through globalised industrial production of animal-based foods has been possible through corporate capitalist systems. Critical approaches to animal studies have also deployed a critique of capitalism in relation to animal oppression (Nibert 2002; Nibert 2017; White 2017) and followed Noske's emphasis on intersectionality (Cudworth 2011). As Richard Twine notes, links can be made between different industrial complexes when it comes to mapping different elements of human relations with nonhuman animals and the interdependence of the crop and vegetable industries with those around livestock and animal feed (2012, 19). While there are many companies which manufacture pet foods, the industry is dominated by five of these: Nestlé (with Purina Pet Care brands), Mars, Proctor & Gamble (Iams), Colgate (Hills Science Diet) and Del Monte (Nestle 2008, 43). Most pet foods have complicated lists of ingredients including, along with meat, fish and a range of animal derived products, thickeners and stabilisers, vegetables and cereals and a host of other additives to improve colour, flavour and texture. Commercial pet foods are tested on animals,[2] often unwanted and surrendered to dog pounds, or dogs bred specifically for use in testing, such as the beagles[3] bred at the Marshall BioRescources puppy farm in Cambridgeshire, in the UK.[4]

The ingredients of pet foods, their journey 'from farm to bowl' and the animal testing involved with most large corporate enterprises are obscured by the packaging and advertising. Rather, advertising campaigns specifically target positive relations of care and concern that human owners have for the cats and dogs they assume responsibility for feeding. In the United States, a Proctor & Gamble campaign for its premium Iams range brought this to the fore with its 'Keep Love Strong' campaign:

> The unconditional love between people and their pets is at the center of the new Iams marketing campaign—Keep Love Strong . . . showcasing the important role premium nutrition like Iams plays in keeping a dog or cat's body as strong as their love.
>
> 'Every day we're inspired by the stories we hear from our consumers about the love they have for their pets and the important role Iams plays in those relationships,' said Ondrea Francy, Iams general manager. 'One of the most exciting things about our new campaign is that it was all inspired by real stories of unconditional love'. (Procter & Gamble press release, 2012)

Chad Lavin (2013) introduces the concept of 'digestive subjectivity' in order to think about the politics of consumerism and public health, understandings of political identity, the operation of political power and the processes and ideologies associated with modernity, democracy and globalisation. Contemporary discourses promoting pet foods can be understood in Lavin's terms of an emphasis on individual responsibility and a shifting of political terrain towards the fashioning of 'entrepreneurial subjects' who engage in self-surveillance. In the case of pet foods, the owner enacts surveillance in terms of their own identity (as a good, loving or responsible pet owner) and in terms of selecting appropriately from an ever-expanding range of products. Food is highly political, and contemporary food discourse is fraught with different kinds of anxieties—around health, weight, ethics or identity, in Lanvin's account. These anxieties are bound up with those around globalisation, wherein many people feel that their lives (and the foods they consume) are beyond their control. As such, we have become, certainly in the West, 'digestive subjects' fraught with anxieties about the embodied vulnerability that attends our consumption of food. At one level, a range of anxieties are associated with the food we eat—we are threatened by the presence of pesticides, trans-fats, salt and super-sized portions. Such anxieties can translate to the feeding of dogs. Currently, veterinarians are concerned with an epidemic of cat and dog obesity which the pet food industry claims will be key to future developments (UK Pet Food 2023, 2). Also political is the production of food within networks of globalised agribusiness that undermines the link between diet and national identity and, potentially, our safety in long supply chains with limited accountability. In the case of pet foods, some companies are keen to market their products as using nationally sourced ingredients as well as focusing on close relationships between dogs, cats and their people. Owners are responsible for the health and welfare of the animals in their care, and care is practically and emotionally demonstrated by feeding. But what happens, for example, when Iams does not keep cat and dog bodies 'as strong' as their owners' 'love'?

Anxieties in relation to the 'politics of pet food' have been considerably heightened since the pet food recall scandal of 2007 in the United States. For two months, from March 2007, a significant number of leading dog and cat food manufacturers recalled products from retailers. The products recalled included familiar premium brands like Eukanuba, Iams, Hill's Prescription and Royal Canin. This was the largest recall of consumer products in US history. In the aftermath, large numbers of canine and feline animal companions (four to five thousand) were found to have died from severe kidney failure resultant from a toxic chemical cocktail present in their food. Contamination had been traced back through the supply chain revealing connections in the sources for pet food and products fed to agricultural animals including the huge numbers of animals grown for meat, milk and eggs to be consumed by human beings. In *Pet Food Politics* (2008), Marion Nestle provides an account of the complicated and unsavory processes which led to the recall, tracing the path of the poisoned food back through the 'interconnected, multinational system of food production, distribution and consumption'. The food industry routinely uses cost-cutting initiatives and, in recent history, adding waste products from one kind of industrial process, such as melamine manufacture, to boost nitrogen levels in animal feed, baby milk and children's food products, for example, has been common. However, it was the effect of combining melamine with cyanuric acid, another nitrogen-rich, cheap chemical, which produced crystals in animals' urine and precipitated rapid and fatal kidney damage (Nestle 2008, 27–52). The outbreak was caused by adulterated wheat gluten and rice-protein concentrate from China, resulting in the execution of Zeng Xiaoyu, fraudulent businessman and former head of the state food and drug administration (Nestle 2008, 127). In the United States, pet owners jammed veterinary surgeries and phone lines, set up internet sites and ultimately a Good Pet Food movement while pet food sales fell by 30 per cent and animal advocates were attempting to 'take pet liability damages into the stratospheric heights of compensation for human liability' (2008, 162). The company at fault, Menu Foods, agreed to compensate owners of dead animal companions for their losses, in terms of the financial costs they incurred (of veterinary bills, cost of purchase and replacement of the animal and so on) (Torres 2007, 59).

While legally in the United States animals are property and might simply be replaced like a damaged piece of household furniture, that is clearly not how they were viewed by those who lost the dogs and cats of their hearts to kidney failure, and in some (few) jurisdictions, such as Washington, DC, successful claims were made including compensation for 'emotional distress'. Another response to the recall was an increase in vegetarian pet food, and a more powerful voice for the raw food and natural diet groups often lobbying consumers via websites on health and the feeding of animal companions. The impact of the recall was felt far beyond the United States and revealed both ambiguities and tensions in our relations with other species and with the foods they, and often we, eat. This is the current trajectory in pet food development with increased concern with the health benefits (or lack of these) in relation to pet food and the need for sustainable sources of ingredients and packaging.

There are certainly alternatives to meat-based diets for animal companions. In the UK, vegan veterinarian Andrew Knight has been a pioneering force, drawing together research on the health benefits of meat-free diets for dogs (such as Brown et al. 2009) and defending veg(etari)an diets of animal companions as healthier (Knight 2005; 2008). The website vegepets promotes such diets, promoting suppliers and brands (see www.vegepets.com). However, this is a limited commercial presence in the UK. Writing with the interests of the animal-based food industries in mind, one of the conclusions made by Nestle was that it was necessary, economically, to recycle animal and other by-products for pet food. Certainly, public distrust of animal-based pet food appears to have been very temporary. Iams, along with other brands involved in the recall, have recovered and expanded their markets, demonstrating also that although faith in the production of animal-based foods for companion animals was undermined, in the long term, belief in the necessity of animal-based foods for companion animals was not. The next section considers the idea of carnism as an explanation for why that might be.

CARNISM AND ANIMAL COMPANIONS

As Melanie Joy points out, all societies demarcate species that cannot be eaten and those which can (2010, 13). These differ across cultures so that while in India both dogs and cattle live either domesticate or feral existence, it is cattle in Hindu-dominated states that are protected and may not be eaten. In the West, however, it is common practice to feed cattle-meat to dogs kept as companions in the home. A common Western response to the idea of eating dog-meat is disgust, but what is of most significance here, Joy challenges, is that most people do not experience disgust at the thought of eating the very small range of animals which Western cultures deem edible. We do not experience disgust because meat eating is an established social practice and normative in our culture. We believe it is legitimate—both ethical and appropriate—to eat animals such as cows, pigs and sheep, and this constitutes a belief system she calls 'carnism' (2010, 29).

Carnist belief systems are predicated on a series of myths revolving around what Joy terms the 'three Ns' of justification: that eating meat is normal, natural and necessary. These ideas are so ingrained in our consciousness that we do not have to think about them—we have internalised them as truths (2010, 95–97). Professionals such as veterinarians, doctors and nongovernment organisations supporting corporate agribusiness represent institutionalised carnism and function as 'emissaries for [carnist] myths' whilst the mass media reinforce the carnist message of 'the way things are' (2010, 100, 105). While conformity is culturally rewarded, deviance is punished and thus 'vegetarians often find themselves having to defend their choices, explain their diet, and apologise for inconveniencing others. They are called hypocrites if they wear leather, purists or extremists if they don't' (Joy 2010, 106). Joy makes the simple but important point that in a carnist society where meat eating is customary and 'traditional', its longevity as an established cultural practice makes it easier to justify and far more difficult to question. A second key form of justification is naturalisation. Practices such as meat eating are seen to be in line with the laws of nature and having a biological basis (2010, 107–8). Third, and closely connected, is the belief that eating meat is necessary, particularly for health, but also for economic growth and stability (2010, 109–13).

Joy considers carnism to be a system whose beliefs we internalise. In the process, the animals eaten as meat are objectified and de-individualised. We also categorise animals into those we are prepared to eat and not prepared to eat based on traditional practices and on ideas about appearance of animals (there are those who will not eat rabbits but will eat turkeys) or intelligence (those who won't eat pigs or dolphins but will eat sheep) or those animals often considered disgusting (rats, snakes, pigeons) (2010, 114–28). What is needed, in Joy's view, is to (re)awaken empathy for animals and invigorate disgust for the meat offered on a plate. However, Joy underestimates the ability of systemic social relations to reproduce themselves here. It is not enough to crack the carnist system and assume that once empathy and disgust are engendered, a moral compass in which all animal products are eschewed follows. Rather, the questions of boundary drawing are more complicated. Erica Fudge (2008, 18–19) emphasises this reinscribing of boundaries in terms of what species of animal are and are not eaten. For vegans, boundaries are clearest of all, and for vegetarians they are reasonably clear in that 'beings with animate life' are 'not to be eaten'. It is meat eaters who live, she says, with boundary confusion, continually making decisions about 'where the boundary between the edible and inedible exists'. Both Joy and Fudge do not consider, however, how the boundaries are blurred for some vegetarians, and even vegans, when faced with the issue of what to feed the dogs (and sometimes cats) with whom they live. In her meticulous Foucauldian account of the 'persistence of meat', Paula Arcari argues that meat persists despite the evidence of its deleterious impact on human health, animal welfare and the environment because we have seen the greenwashing and 'ethicalisation' of the mainstream meat industry (Arcari 2020, 39). The questions of who eats what mammals, birds and fish, are problematic, particularly for owners who live with dogs (biologically classified as carnivores, but often seen as omnivores with a physical and undeniable carnivorous bias) and cats (classified as carnivores and often seen as obligate carnivores). As we will see in the second half of this chapter, feeding animals does produce ethical tensions, and the tension for vegetarian and vegan dog owners is likely to be exacerbated by the persistence of carnism by the pet food industry and in their advertising, alongside moves towards the ethicalisation of pet food.

Pet food advertising involves the use of animals, which have been shown to boost sales for products, generally, and for pet food, specifically (see Lacendorfer, Aitken and Reece 2008). In such advertising, animal actors make their preferences clear, often addressing human owners/consumers. In an Iams cat food television advertising campaign, the feline actor is blunt: 'I am not a vegetarian', says the voice-over, 'these teeth were made for meat' while for the next cat on screen, its appearing to be hunting is voiced-over with 'Do I look like I'm stalking plants?' (Iams 2015). This advertisement ends with a cat demanding to be fed 'what I'm born to eat'. The consumption of this food is thereby naturalised for animal companions. Similarly, pet food advertising plays on the naturalness of the meat-based diet as something that will satisfy a dogs' instinct or nature. In promoting Crave brand products, for example, the eyes on the packet are not of a dog, but of a wolf. The Crave television advert of 2019 in the UK had the tagline 'Respect the Animal Inside' and pictured a variety of sizes and varieties of dogs and cats walking towards the camera with a focus first on a cat accompanied by the words 'born hunters' and then on a dog with 'not gatherers', in both cases, the camera zooming in on feet and claws. Lest this did not sufficiently clinch the idea of the naturalness of a meat diet, there is a final message encouraging the potential consumer to 'satisfy their nature' by feeding cats and dogs their 'high-protein', high-meat content foods.

In similar vein, the advertising for Canagan brand pet food enhances this narrative of primal instinct by claiming that their products are a 'natural diet based on the Ancestral Ratio™ of prime meat to carbohydrates wolves and wild cats would hunt in the wild' (www.canagan.com/uk/), whatever an ancestral ratio might be, however, is left to the imagination. The logo accompanying the narrative depicts the head of a wolf-like dog baying, presumably at the moon. The link between the logo and the product is clear, for designer Chris Mitchell, who produced it:

> The recipe for Canagan pet food is a nutritionally balanced combination of premium meat and carbohydrates in a grain-free format that mimics the diet of the canines wolf ancestors. The illustrated Canagan dog icon had to appear wolf-like in character to reflect the evolution of domestic dogs from their wolf ancestry.[5]

While the pet food industries emerged from Western countries (and indeed are embedded in the colonial history of the global meat industry, see Cudworth 2011a), other producers who have been latecomers to the pet food industry deploy similar narratives of the naturalness of companion animals eating other animals. In South Africa, the key pet food producer in the continent, apartheid meant very limited market development until the 1990s. However, some companies have focused on novel animal-derived ingredients for pet food, such as deer meat, warthog meat and meat from free-range rabbits and ostriches. This represents a new twist in the naturalisation of carnism, linking this to sustainability. In presenting the novel ingredients of the food made by Maneli Pets, its founder and director asserts that meat from such animals is 'free range' and 'sustainable, because these animals are culled for biodiversity purposes' (Sipra 2022). This represents a strong example of the ethicalisation of meat as pet food. In marketing the product as both natural and intrinsically South African, the company makes the startling claim that they are setting out to 'replicate what cats and dogs ate before human intervention in the African savannah'.

The recycling of animal parts as pet food has become particularly marked in the 'treat' market, where parts of animals that would not be eaten are sold as healthy and natural supplementary foods. For dogs, a quick browse reveals options which are apparently more mundane such as pigs' and goats' ears, the tails, tendons, ligaments and penises of cattle or the hooves of calves, lambs' lungs, dried whole fishes or fishes' skin. Alternatively, more exotic choices might involve the tendons, trachea, knee joints or skin of ostriches. In addition to various animals' bones, a Golden Moonbone can be found, composed of 100 per cent 'dried beef cartilage' which is 'air-dried to perfection' and grain free, super low fat, high in protein and suitable for puppies. JR Pet products recommends their primal range of dog treats, where the ears of various creatures (deer, rabbits, cattle and so on) retain the animals' hair, providing:

> many benefits . . . as products in this form are as close as it gets to what our canine friends would have eaten as wild animals. The hair acts as a natural wormer and helps maintain dental hygiene.[6]

Another means by which meat-based food is naturalised and normalised for dog companions is by being presented as something that human purchasers can recognise as food, and which dogs themselves would choose. Led by Pedigree Pet foods in the United States, and their 'We're for dogs' campaign (the longest single campaign by a pet food brand), dog advertising has become increasingly dog-focused. Butchers (2021) recent advertisements are focused around the theme of 'Nourishing food for dogs', telling us that a dog is 'more than man's best friend' but also their best 'work mate', 'play mate' and 'soulmate'. Ergo, their reward should be 'a nourishing, wholesome bowlful' of something largely composed of meat, as the brand name boasts. Similarly, Harringtons (2021) tells us that 'everything comes naturally to dogs' who are apparently 'natural companions', 'natural huggers' and more, and who deserve 'food that's natural, wholesome and tasty' and meat based. Here, the naturalisation of the species-story of dogs which we encountered at the start of this book segues with the naturalisation of meat from farmed animals as food for dogs. A further nutritional motif is the presentation of food for dogs in ways that are similar to food for humans. 'It's like having my own personal chef', says the voice-over for a dog in the Tails.com (2021) advertisement for personalised recipes which can be ordered online. Dog food is also increasingly produced in ways that mean it looks closer to human food. The packing for Marks and Spencer's 'Complete' premium dog food, for example, pictures meat and vegetables that might be found on a human dinner plate. Interesting in this case, is the erasure of the animals who become dog food as '100% British and Irish meaty goodness' on the packaging. The post-recall concerns with traceability are also present in the nationalist framing.

In advertisements for Butternut Box (2022) mail-order dog food, we are told that 'dog food through the ages has been a bit underwhelming' because 'it's never really looked like food'. While their range does feature one vegetarian and one vegan wet food, the overwhelming majority of products are meat based. The humanisation of the food is achieved by recommending the products as 'very *real* food for your dog' and the website makes clear that this food is 'tested on humans' (Butternut Box 2023). Drawing on dog owners' relationships with their dogs, and their desire to provide them appropriate care, these are powerful narratives that shape dominant ideas on what it is appropriate for dogs to eat. As animals sell products and can be legitimately used in such advertising, pet food adverts abound, and advertisers are finding ever new and ingenious ways to sell their products. Brand market leader Pedigree, for example, has recently found an innovative way to target consumers (and circumvent GDPR[7] restrictions) in advertising to them by using dog recognition technology on smart billboards such that dog food advertising appears when people are walking with their dogs in urban areas (Contagious 2021).

The remainder of this chapter considers how carnist ideas might be seen to underpin our processes of categorising nonhuman animals and impact what is eaten and how it is eaten in multispecies homes. It looks at the ways in which these categorisations might be disturbed and sometimes reconfigured by the practice of living with a dog companion. In interviews focusing on their lives with dogs, many people discussed the food they fed their dogs and often also, this spilled over into talk of what they ate themselves. The majority of interviewees were 'carnists', but many of these placed restrictions on the range of animals they would eat, and many were also concerned for the welfare of animals other than dogs. In this sense, they are dealing with the problematic boundaries of the animals we may and may not eat, and for many, this was troubling. Most interesting were the vegetarian and (lone) vegan dog owners for whom one might imagine the carnist glass to have been cracked. However, in every case, dogs and cats in the home were fed meat. Carnism reinscribes itself in the ways in which veg(etari)ans living with dogs discuss the ambiguities, tensions and reconciliations between their belief systems and the practice of feeding meat to the dogs-of-their-heart. Since this data was collected, however, change is recently but firmly afoot, and the pet food industry is charged with sustainable accountability with pressure mounting year-on-year. This chapter will close therefore with a consideration of the possibilities of plant-based diets for dog companions.

FOOD AND EATING IN MULTISPECIES HOMES

There were, surprisingly, no substantial differences in data obtained from interviewees living in urban and rural locations on what people considered that the dogs who lived with them might eat. One difference of note, however, is that the vegetarians were more strongly represented amongst the London-based interviewees. In the course of our discussions, negative views were often expressed about industrial animal agriculture ('factory farming') and industrial models of pet production ('puppy farming'). I would suggest, therefore, that when it comes to food and eating in multispecies households there is both ambivalence and a reinscribing of species boundaries (between dogs and farmed animals) *and*, although to a lesser degree, a broadening of concern with issues of animal welfare. In the quoted material below, and for this chapter only, I have chosen to differentiate between interviewees as those meat eaters who place some restrictions on what they will eat (restricted carnists) and those who do not (unrestricted carnists) in addition to those who described themselves as vegetarian or vegan.

Who Can Eat Whom?

A question often asked of companion animal ownership is whether it enhances or augments a wider concern with animal welfare or even an understanding of animal rights, or whether it places lovers of dogs in a situation of ambivalence as they continue to eat other animals. For some who live with dogs, and for whom eating meat is ethically unproblematic, discussions of what companion dogs eat reveal no sense of anxiety around species boundaries. In this quote below, the anxiety focuses on which food will not spark an allergic reaction rather than what or whom is being eaten:

> Everything's got to be fish based, so we've got him on Chappie and it took a lot of time to find one that didn't flare his ears and his eyes up. The vet suggested we put him on Hills, which sent him hyperactive—sixty quid[8] a bag, as well. It sent him hyperactive, so I whisked him off of that. We also had one called Wafcol, but it made his ears quite bad so we experimented, and we went down to [a large pet store]. We get their own brand of the fish and rice, the dried food. So he 'as that and Chappie. He can have pigs' ears for treats, pork strips for treats, anything pork or ham he's fine with. (Unrestricted carnist)

However, issues of animal welfare did surface in many of the interviews, in particular around the use of animals as food. Where this was raised as an issue by interviewees, it was in terms of their own vegetarianism or the expression of conflicted feelings around eating animals, and in many cases, this was linked to ideas about people's relationships with their companion dogs. In another case of a dog with a restricted diet because of allergies, there is a very different narrative to that of the interviewee above, this time from a vegetarian owner:

> It just upsets me, so I have not to think about it because it's just awful. We've tried everything and it's hopeless. We've done all the kinds of vegetarian food, we tried lots of different fish, fish and rice-based ones and that all made him really ill. He can't eat beef, he can't eat chicken, it has to be pork or he is ill. And that's just about the worse thing, isn't it? It's just the worst thing, and god knows how it's produced. Well, we do know, don't we? In factory farms, awful. So, Mungo eats pigs and I hate it. I hate it. (Vegetarian)

Clearly here, there is a hierarchy implied between species in terms of the least poor ethical choices that this owner might make. Despite the tension in the narrative, any dilemma is resolved in favour of the health of the dog rather than 'make him ill'. Important to note also is that this interviewee considered that the least ethical option would be to give her dog up for rehoming so that 'someone else gets to buy all that stuff and not me'.

In another case, a similar hierarchy of least poor ethical choices is both resisted and legitimated by recourse to a rather odd form of naturalisation:

> I don't think it is more acceptable for a chicken to be slaughtered than a cow, but somehow, I can, I can handle the thought of Rocky, as a dog, eating a bird more than I can handle the thought of him eating a lamb or a cow because he's so small, I think. I don't know, but there's something there. I don't know what the logic is but there is, there's something that always sends me to the chicken-based thing [dog food]. (Vegetarian)

It is not only vegetarians for whom the boundaries between different animals and what may/not be eaten become problematic. In this discussion between two interviewees (a couple), the questions of species and boundaries are brought into sharp relief when they move from talking about what their dogs eat, to what they eat, with disgust, naturalisation, denial and anxiety all making their presence felt:

> Well, I do eat other meat [than chicken meat] but I do find it hard not to be sick sometimes when I think of what you're eating. I haven't got any conscience where chicken's concerned. It's horrible really.
>
> [Partner interjects] I'd be more concerned about eating chicken, because of the welfare of the chickens rather than cows. I mean, I'm not an expert, but I imagine them to be romping round the fields, before, you know, at least for a bit, you know, more than your average factory chicken . . . I mean it's horrible, isn't it?

> I mean, we're not vegetarian but we don't eat much meat that's not white meat. Although, you know, I don't think it's necessarily wrong to eat meat and, um, I've had rabbit and, you know, other things that may have had a better quality of life, but I suppose I'm like a lot of people, I just push it out of my mind all the time. And pork is probably, well, although it's one of my favourite meats, it's probably the one that troubles me the most because they're supposed to be as intelligent as dogs and have similar emotions . . . which is quite chilling really. (Restricted carnists)

For a number of interviews, discussions of everyday lives and relations with dogs leads them into a comment on their own dietary practices and, often, feelings of guilt are expressed about eating meat:

> I carry it around all the time, the guilt, looking for organic and free range. I don't eat much meat, and when I do, I feel guilty. I should give it up, really, it's bloody ridiculous. (Restricted carnist)

Or alternatively, a general love of 'animals' or 'wildlife' raises this tension:

> When I go to the allotment, I usually leave something out for the fox, especially if it's wintertime. I don't know, I just love being around animals, and I do feel guilty actually, that I eat them. It's a bit of a quandary. There's a little robin that comes and gets worms when I'm digging. (Restricted carnist)

All the dogs that interviewees lived with ate meat, however, and this was both normalised and naturalised with a form of biologic essentialism advanced to justify a diet of meat for dogs.

'I'm a Vegetarian, but She's a Dog'

A common sentiment expressed by those interviewees who did not eat meat was a version of 'I'm a vegetarian, but s/he's a dog'. Also important, therefore, is the notion that human beings are different from other animals in their ability to exercise choice over what to eat and what not to eat:

> It's not a conflict, I don't mind getting meat for her, she's a dog so I don't see that as a problem. I don't mind other people eating meat—if they are eating meat, it's their choice. (Vegetarian)

> the thing about vegetarianism is that it's to do with humans having a choice and not being under any necessity of eating meat. Dogs are essentially wolves and it's an essential part of their diet. (Vegetarian)

Here, the species story of the dog as the wolf descendant and thereby assumed to be an obligate carnivore comes across strongly, and this was a common assertion, perhaps picking up some of the queues common in pet food advertising. One-quarter of the interview sample were vegetarian and a number made a direct connection between their diet and living with a dog:

> But I do think that living with an animal does raise interesting issues about eating. I mean for me, I find it hard to believe that lots of people live with animals but quite happily eat other animals too. To me, being that close to an animal [his dog] sort of reinforces the idea that for us, for humans, we have relationships with other animals or we're capable of having relationships with other animals which are different to those of some other animals. I know loads of people who do live with animals either as pets or as farmers, and happily eat them, but certainly with me, it reinforces my natural inclination to stick to grass. (Vegetarian)

> I know loads of people who absolutely adore, love animals, and they are very bothered about animal welfare but they eat meat. But . . . you always think that thing of 'I could not eat my animal, they're animals and I couldn't eat my dog', I just couldn't . . . because you make that connection, that's an animal that you are eating, I think, I mean lots of people just put it to the back of their mind probably, but when you're biting into that meat, I just imagine that it's Sam—yuk. (Vegan)

For vegetarians, feeding meat to dogs was discussed through a tension between the carnist assumptions that dogs needed to eat meat, alongside vegetarian repulsion at what (or whom?) was being decanted from the can into the bowl:

> Well, I can cope with it, you know, just about. I wasn't brought up as a vegetarian as a child but [name of partner] was, and he can't really cope with meat at all. I know when he's feeding the dog because I can hear him heaving in the kitchen. (Vegetarian)

> I do give him like dried dog food, but I can't cope with the kind of canned sort of meat thing, and I think it is better for him to have, like, proper food rather than out of a tin, but when he's with my ex-partner—he has him two days a week—then he gets tinned dog food. (Vegetarian)

For those who cooked food for their dogs in addition to or as an alternative to commercial dog food, or who fed dogs a raw food meat-based diet, there were further issues. For vegetarians, this meant buying and cooking meat they would not otherwise buy, and occupying spaces they would want to avoid such as butchers' shops in search of cheaper or better-quality meat. Cooking meat raised particular conflicts where normative and natural discourses of carnism are brought into proximity with vegetarian disgust:

> I don't actually buy free-range chicken, I buy just bog-standard chicken. That goes against all my principles, not buying free-range chicken. One thing is buying the meat in the first place and then I'm not even buying free-range chicken. You know, there's only so far you can go. . . . The stuff I buy her is chicken breast, 'cos although I could get it cheaper if it was on the bone—I'm prepared to cut up chicken, but I do it with a fork and scissors, but I don't touch it with my hands—but there's no way I'm going to spend time pulling or cutting meat off bones. (Vegetarian)

> I have to limit the amount of meat I can give them just because I can't pick the chicken off the bone. (Vegetarian)

For carnists, the cooking and preparation of meat for dogs presents no such challenges and this is far more readily integrated into daily routines and reflects the desire for the food dogs eat to be human-like:

> every week they have a cooked chicken. They don't eat the bones or anything like that . . . we hand strip the chicken and they have that, we cook lamb mince as well, so you know, and obviously if we're eating, we'll give them something, but it will only be [a small amount]. I'll try and put vegetables in their food, put gravy in it just to give them a bit of vitamin and stuff like that so, erm, you know, they have their own type of food . . . I adapt our human food to suit their diet, really. (Carnist)

Vegetarian disgust is overcome by a combination of carnist discourse of naturalisation, necessity and normalisation, combined with an inflection of dogs as active decision-makers in the process, choosing to eat meat. In some cases, this is based on the behaviour of dogs when presented with a variety of foods:

> I wouldn't ask them to be vegetarian, they really do enjoy their meat, and I, just because I choose to be vegetarian, there's no reason why I expect my dogs to be vegetarian and they love their meat, absolutely adore it . . . but I would like, ideally, I would like them to have, erm, organic chicken, organic meat, plus a reasonable amount of vegetables, but they're very naughty on eating vegetables . . . their choice is not to eat vegetables, be vegetarian and therefore just because I choose to be vegetarian there is no way I would ask my dogs to be vegetarian. That's my choice, I'm not gonna impose my choices on them. (Vegetarian)

> having thought about it, I decided that I didn't think that decision was really mine, mine to make that I should enforce my dog to have a meat-free diet . . . I spoke to my friend who was a vet and he was very against the idea of him being

> a vegetarian and I decided actually that is quite, might be quite a, erm, controlling thing for me to do. My dog will be vegetarian when he clearly really likes meat, yeah, erm, yeah, we've tried we've given him vegetarian treats ages ago and he'd just go, you know, he wasn't interested so, so no, he's, he's [pause] he will always eat meat. (Vegetarian)

Ultimately, carnism and the notion that dogs have a strong preference for eating other animals overcomes vegetarian dog owners' conflicted and sometimes very troubled feelings. For most, there is the notion that for dogs, eating meat is 'meant to be' despite many acknowledgements of the problems of commercial pet foods (on both health and ethical grounds) and the dilemmas faced by vegetarians in preparing/cooking meat for canine companions:

> There is, there's a contradiction, there's lots of stuff, y'unno I, there's lots of literature about vegan diets for dogs but I'm not, I don't know [pause] I'm not convinced by it at all. I don't know about a vegetarian dog . . . they're not the same as us, they don't have the same enzymes . . . I don't even think if they are always vegetarian [from a puppy] it would work, . . . It's like trying to give a horse [some] meat or a sheep a nice bit of steak, I think there some things that are meant to be. 'You like your meat, don't you?' [directed at the dog]. (Vegan)

> At first, I found it quite hard, but I suppose I make a compromise and, like, I don't, I tend to buy organic, 'happy' chicken, they had a happy life, you know, even if it was very short and it makes me feel a little bit better about it. But I kind of accept it, you know, he's a dog and he's not, he's not a vegetarian, erm, and I think the whole, you know, the raw meat and the bone and stuff is really good, good for him. So I think, I just think his health is more important really, so, so he gets well fed and never gets fat. (Vegetarian)

Certainly, all the dog owners I spoke with were concerned with the health of their animal and considered that this must be a priority. In some cases, the dissemination of carnism was reenforced through the advice of veterinarians. Some of the vegetarians interviewed made specific points about pet food corporations from whom they would not buy 'because they do horrible things to the animals they test in their labs'. Yet for one vegetarian interviewee, the opinion of an expert overrides this critical knowledge: 'I buy Iams because our friend that's a vet says that's the very best food you can buy for them because it's already nutritionally balanced'.

CONTESTED CARNISM AND PLANT-BASED POOCHES

The responsible dog owners in this study cared very much about the health, welfare and happiness of the dogs with whom they shared their lives. They had considerable amounts of what Leslie Irvine (2004, 66) calls 'animal capital'—resources that enable non-exploitative relations with animal companions such as knowledge about training and health or an active interest in animal emotions or cognition. The interviewees also had concerns for animals other than their own dogs, certainly for dogs in general, for animals kept as companions more widely, for wild animals and in some cases, for non-companionate domesticate animals. For the 'unrestricted carnists', issues of an appropriate diet for their dogs were unproblematic, but the majority of people interviewed did restrict their carnism in certain ways and their discussions of their own diets and that of the dogs they live with reveal tensions and uncertainties regarding what animals can and should be eaten as food. For the vegetarians, ambivalence was heightened.

All the vegetarians with whom I spoke claimed to 'love animals' and this, often alongside a history of having close relationships with dogs as companions, was given as a reason for their own choice of diet. All of them bought and fed meat—sometimes that they had prepared/cooked themselves—to the dogs in their care. The 'three Ns' of carnism make their presence felt in discussions of animal diet. Feeding dogs meat is seen as essential for their well-being (in the case of those with dietary allergies) or more broadly in terms of the most nutritious and healthy diets possible. Despite rejections of carnism in their own diets, these interviewees normalised and naturalised a meat diet for dogs by drawing on arguments for biological difference and physiological necessity. As the lone vegan in the study put it: 'Some things are just meant to be'.

My vegetarian interviewees were not naïve—many had considered a vegetarian or vegan diet for their dogs, and some had experimented with such diets before deciding to reject them. Clearly others of them felt the pressures of carnist normalcy and felt that there was no other decision to be made than feeding their dogs a meat-based diet. There are dogs (and cats) living well on vegetarian and vegan diets, and research evidencing the suitability of vegan diets for dogs (Knight et al. 2022), despite the carnist responses of veterinarians and some charities concerned with the welfare of companion dogs. There are well-made arguments suggesting that just as we humans might choose a vegetarian or vegan diet for ethical, environmental or health reasons, the same process of reasoning can convincingly apply to a meat-free diet for dog and cat companions (Knight and Leitsberger 2016). Yet such arguments and, in some cases, experimentation with vegetarian/vegan dog diets failed to convince the vegetarians in my study that avoiding animal products resulted in an optimal diet for the dog in their care.

While the implications of a culture of carnism are clear here ('dogs are essentially wolves'), I also think that the quasi-human quality of dogs themselves has a part to play. The normality of pet dogs eating meat means that feeding otherwise raises anxieties about health for those that might otherwise be drawn towards a veg(etari)an diet for their companion animal. Certainly, this has been supported by recent research on owners' views and plant-based diets for pets, where those avoiding meat in their own diets were found to be disproportionately represented among those living with animal companions compared to the general population, and thereby likely to experience the dilemma of their dogs eating other animals. While all owners in Susan Dodd et al.'s (2022) study, for example, were found to express concern for farmed animal welfare, it was only vegans who fed plant-based diets to the animals in their care, and these were only those vegans who did not see a plant-based diet as 'unnatural' or as a health concern for their animal companion. The vegetarian interviewees in my study were those who expressed the highest levels of ambivalence about the keeping of dogs as pets and were concerned about their lack of freedom. Whilst contradictory, therefore, perhaps it is the desire to enable an element of choice in the lives of their dogs that such dog owners would make reference to the preferences or choice that their dogs might make in favour of meat.

Given the obvious affection and love in relational ties between people and animal companions, the carnist culture of pet food advertising is pernicious in suggesting that those who care for animal companions are obligated to feed those they love other, dead, animals. The culture of carnism, then, inscribes who (or 'what') is killable (Wolfe 2013), who can be made to live and made to die, and for whom. We have, then, an embedded, lucrative A-IC legitimating its existence through a range of cultural tropes so powerful that even those who question some elements of the A-IC, in terms of their own diet, may be implicated in others. There are certainly fervently held views on what it is appropriate to feed dog companions. In the many posts to be found in response to articles on the web, there are some strongly held views. There are those who express great clarity. This includes the 'vegan view' that keeping dogs as pets is unnatural, commercial pet food is unnatural, ergo there is no 'natural' diet for a pet dog and dogs can happily be vegan. For others, dogs are carnivores and must be fed as natural a diet as possible, thus raw food, including large quantities of flesh, is optimal. Many are convinced by 'science' and 'expert knowledge' that suggests established 'pet foods' are both the safest and most reliable option, and certainly, the cultural tropes of pet-keeping reinforce this strongly. In the field, and the parks and the kitchens where dogs and their people muddle along, there is often ambiguity and a lack of clarity. As such, even those who may contest carnism in their own diet and many aspects of their daily lives just 'don't know about a vegetarian dog'. In what Haraway (2008) calls the 'sticky knots' of our interactions and co-dependencies with nonhuman animals, ambiguity is inevitable. While I sympathised more with some of my interviewees' attempts at ethical relating than others, most were very much concerned with trying to live well, or at least to live better, with some other animals. And that rather muddy world of ethical relating is something in which I chose to take heart.

This said, given that the UK pet food industry is challenged to respond to criticisms that it is unsustainable and needs to consider the welfare implications of its operation (in terms of animal testing and the lives of farmed animals) (UK Pet Food 2023), I hope such pressure might lead to increased opening for companies producing plant-based dog food and the shift to plant-based becoming mainstream. This has begun with mainstream producers offering plant-based options (see figure 4.2). In her recommendations for undoing the 'cartographies of meat', Arcari sees the breaking of associations between meat, 'food animals' and consumption practices with popular understandings of nature and naturalness as crucial (2020, 291). With pet food, this will be a challenge, given, as we have seen in this chapter, the way dogs' species story shapes how food is promoted for dogs who are 'born to hunt' and to eat meat.

Figure 4.2 Barking Heads plant-based kibble.

Pure pet food castigated dog owners for feeding dogs 'highly processed brown biscuits'. A personally tailored, mail-order 'fresh' food diet alternative to the brown biscuit[9] was recommended as food that owners who care should feed their dogs because 'dog owners treat their dogs like their kids' (Pure 2020). This, as we have seen, is a key trope of dog food advertising and relies on another element of the species story of dogs—that they are 'natural companions' who enjoy human company and being handled and who provide us with 'unconditional love'. Both our relations with dog companions and the food dogs are encouraged to consume is naturalised—the good dog owners feeds their dog 'well' because they are 'family'. In the next chapter we will consider to what extent this familialising of dog companions resonates for those living with companion dogs.

Chapter 5

Making Oddkin

In 2017, a judge in Rome ruled that the days away from work that a university librarian took in order to care for her elderly dog should be regarded as an absence due to 'serious or family personal reasons' (BBC 2017). The woman had requested paid leave in order to take her elderly dog to the vet and provide post-operative aftercare, but this was rejected, prompting the legal action (*The Scotsman* 2017). The ground for the successful legal case was that she argued that her dogs were her family, and this was seen as a landmark ruling by animal protection organisations in Italy (Specia and Pianigiani 2017). In the first month following the Russian invasion of Ukraine in February 2022, it was estimated there were 2.5 million Ukrainian refugees, half of whom had fled with their animal companions (Taylor 2022). Research into the role of pets in households of people who remained in the capital, Kyiv, during the conflict indicated that because 'pets are considered family members', many people refused to leave because they could not abandon them. Additionally, animal companions provided a source of comfort and security (Kateryna et al. 2022). Those leaving found that some countries lessened or removed the requirements at borders for pet entry from Ukraine, while media reports include human migrants' descriptions of their animal companions as family (de Ferrer 2022).[1] The idea that pets can be family has popular traction and, as we see in these examples, has had legal and political impact.

This chapter concentrates on the emotional attachments between people and animal companions, focusing on the range and quality of those relations and considering whether dogs are understood to be family members or kin. Some sociologists have suggested that multispecies households are a social phenomenon of sociological interest (Charles 2014), while others have gone further and suggest we have seen the development of multispecies families (Laurent-Simpson 2021). In the West, cultural understandings of families as being interspecies in nature are increasingly normalised in mainstream discourse (Laing 2020b), with survey data indicating that animal companions are considered family members and that the number of people thinking in this way is rising (Harris 2015). Within the sociological literature on animal companions and the family/household there is ambiguity, however. Drawing on the ethnography, this chapter maps the kinds of emotional bonds and affective relations people have with dog companions. In doing so, it problematises the assertion that animals such as dogs are understood to be family by standing in for other kinds of relationships humans may have and/or by being humanised, particularly by being understood or represented as children. It will further argue that the sociological concept of the family is too stretched and burdened to capture intraspecies domestic relations. Through a critical reading of Haraway's conceptions of multispecies kin and kin-making, this chapter explores whether those wanting to conceptualise affective relations between people and animal companions might abandon the notion of family for the more inclusive possibilities of kin. Haraway uses kin-making to indicate a broader intimate relational landscape than relatives of blood or intrahuman intimacy, socially sanctioned and often legally prescribed ('godkin'). Her notion of 'oddkin' means that intimate relationality might be more widely drawn, beyond immediate human relatives to a wider range of people and beyond the human (Haraway 2016a, 2). This chapter brings Haraway's conception of oddkin into conversation with the sociology of family and personal life in order to open up posthumanist possibilities for conceptualising intimate relations.

This chapter begins by outlining the ways mainstream sociology has maintained an exclusively human perspective on the family and personal life, and looks at the ways some sociologists have sought to incorporate animal companions into the study of the family. The second section interrogates the possibilities of kin as an alternative or supplementary concept to that of the family, and the ways Donna Haraway (and others) have developed this to account for intimate relations with the nonhuman. The chapter then turns to data from the ethnography to explore the emotional bonds people have with dogs and how they think about their relationship. The affiliations mapped by the data analysis will then be considered in relation to conceptions of family and kin in the final section.

SOCIOLOGY, THE FAMILY AND ANIMAL COMPANIONS

Work within the sociology of family life has tended to proceed with little or no recognition of nonhuman members of the household. Even recent texts which claim to fully embrace the extent of change and diversity in the family—with new intimacies, diverse relationships and varied living arrangements—have nothing to say of nonhuman animals (Chambers and Gracia

2021). As we have seen, the social sciences have tended to draw a 'hard line' between the human and the animal (Alger and Alger 2003a, 69), and this has rendered animals invisible in most scholarship in family studies, social work, sociology and other disciplines concerned with the family (Laing 2021b).

When it comes to defining family, sociologists deploy different criteria. Tight definitions of family are based on genealogy (blood relations) or legality (relations formalised in law) despite acknowledging a variety of forms (Berger and Carlson 2020, 479). Looser definitions focus on personal relations which can be based on connections other than blood ties and are part of a critical literature which suggests the institution of the family is too narrow to capture contemporary Western practices. For Ulrich Beck and Elisabeth Beck-Gernsheim (1995, 2), marriage and the family are relics of the industrial age that have been replaced by new forms of multiple and alternating relationships. The family has been understood to be in a condition of fragmentation such that we need no longer associate it with a particular institutional form (Smart and Neil 1999). Rather, Carol Smart proposes a sociology of personal life, understood as a set of practices, and conceptualised as varied sets of intimacies and relationships, incorporating diverse sexualities, friendships and acquaintances (2007, 188). Critics suggest that this model of relationality risks 'overstating the open-endedness of intimate relations' while also running the risk of 'obscuring power relations' (Gilding 2010, 757). Gilding has a point, especially when it comes to thinking about the persistence of traditional conceptualisations alongside a multiplicity of institutions which 'deal in some way with ideas of parenthood, kinship and marriage and the expectations and obligations which are associated with these' (Morgan 1996, 11). Smart acknowledges that the concept of family remains a key referent in everyday language and that whatever the domestic formation 'love and support for family members and kin' is provided (2007, 13), while the literature on 'families of choice' has provided a counter to heteronormative understandings of family and looked at ways kinship is alternatively constructed involving chosen members, be they relatives, friends or lovers (Weeks, Donovan and Heaphy 2001). Whether a narrow or loose definition of family is adopted, or a fluid conception of personal relations, the focus remains firmly intrahuman.

Some of these theories about family change have been incorporated into models of changing relations between humans and animals, however. Adrian Franklin (1999), following Beck, has suggested human relations with animals are characterised postmodern conditions of ontological insecurity wherein pets provide security and compensate for a lack of human affection and, as a result, are increasingly becoming 'companionate family members' (Franklin 2007, 138). Andrea Laurent-Simpson (2021) contends that a specific multispecies family form has emerged through increased relations of intimacy, because living with animal companions can be seen to impact extended family interactions, fertility considerations, parent-child relationships, family life course and finance matters and the structure of the household. For example, the level of emotional connection and the meaningful relationships which people have with the animal companions with whom they live have been proposed as an explanation of why US women delay having children until they are older (Laurent-Simpson 2017). Others accept the concept of multispecies family but reject the idea that this is a new, or a posthuman, family form (Charles 2016, 2.4). For families to be understood as posthuman, Charles suggests, the 'power differentials between companion animals and their humans would need to be challenged and the distinction between human and other animals, which reinforces human superiority, questioned' (2016, 6.1). Charles admits that some kinship practices can be seen as posthuman in blurring the human-other animal boundary, yet the persistent power relations along that boundary mean that it is difficult to see that a posthuman family form can be identified empirically (Charles 2016, 9.2). In addition, there is the important notion that while animals can be human-like in their intimacies with family members in the home, at times, they are seen as more-than-human in terms of being 'better at being family' than human members of a household because they provide uncomplicated affection without strings and 'just love you as you are' (Charles 2014, 11).

Charles's rejection of the notion of the posthuman family form is founded on relations of species inequality within multispecies households and the dependency of animal companions such as dogs. Yet children, and adult humans who have disabilities and/or are aging, may also be incorporated into households in relations of dependency. As feminist scholarship and work in childhood studies has shown, the family may be reproductive of binaries, associated power differentials and boundary-reinforcement in terms of adults and children or in relation to gender. Charles's understanding of posthumanism as an abandonment of binaries is also problematic. This is because for many identifying with critical posthumanist positions, the troubling of species boundaries is a sufficient condition for a phenomenon to be described as posthumanist. In addition, critical posthumanism acknowledges persistent human domination (see Cudworth and Hobden 2018). To consider that something may only be described as posthumanist when species inequality is no more is to assume that posthumanism is a utopian future condition, rather than a way of seeing and thinking about species. As a result, I consider the debate on a change in family form to be tangential to a sociological account of the place of nonhuman animals in the family. What is key, in my view, is the impact of the quality of relationships as evidenced in animal companions being surrogates and substitutes for human familial relations. In short, being like-family seems to be key to the case for animals being family.

To this end, dogs have been described as surrogate children, or 'fur babies' (Greenbaume 2004; Turner 2001). It has been suggested that research on US families without children shows that animal companions can fill or surrogate for the role of the child, and that adult humans gain similar emotional and psychological fulfilment from nurturing or caring for a dog or a cat as they do for a child or grandchild (Laurent-Simpson 2021). Research does indicate that some owners, particularly in households without human children, consider themselves to be 'pet parents' and consider dogs and cats to be 'children' (Volsch 2018; 2019). Other qualitative studies of 'interspecies families' found, however, that although individuals may say that their pets are family, not all considered themselves to be parents or to relate to their pets in a parenting role. Rather, there were differences between those with and without human children, with childless people tending to construct narratives of pets-as-children, those with younger children emphasising the differences between pets and children and those with older children noting the similarities (Owens and Grauerholtz 2019). There has been interest in the effects of different 'parenting styles' on the relationships and bonds owners have with their animal companions, as well as on animal behaviour (Cimarelli, Turcsán and Virányi 2017; Volsch and Gray 2016). However, survey research using scales of 'family bondedness' between people and cats and dogs give ambiguous results and suggests caution is needed in estimating the extent to which relations with pets can be seen to be the same as with human family members (Nugent and Daugherty 2022). There is also evidence that people resist the idea that animal companions may provide a substitute for relations with other humans (Charles 2016, 5.4).

Emma Power (2008) contends that there are diverse understandings of dogs as family. In her research using interviews and diaries kept by dog owners, she found three ways in which dogs were incorporated into families. First, some of her participants understood dogs as family because they are similar to children in requiring care. Second, dogs were incorporated in the family understood as a pack, with participants emphasising the inherent 'otherness' of dogs, which meant that they were treated differently to human family members (Power 2008, 552). Third, the agency of dogs was recognised as shaping family and home, as dogs actively participate as family members in the everyday practices of home (2008, 536). In these second and third understandings, human family members accommodated the 'dogginess of canine companions' and resisted seeing dogs as 'furry humans' (2008, 541). Power argues for a broadening of the concept of family beyond biological relations to include more-than-human relationships forged through cohabitation and everyday interaction in shared space. My reading of Power, however, is that the concept of family is being considerably stretched in attempting to accommodate the second and third ways in which she suggests that dogs are integrated into human homes.

Species inequality certainly impacts household relations. The power of humans to determine the quality of dog lives is significant, and dogs are vulnerable to the whims of human owners and may be caught up in dysfunctional human lives. As we have already seen earlier in this book, legally, dogs are property; commodities that are forced into breeding and sterilisation, bought and sold, abandoned to shelters and even killed when they no longer fit their humans' needs or beliefs about how dogs should behave (Boyd et al. 2018). Where domestic violence and abuse is present, animals tend to also be victims of violence (Fraser and Taylor 2019); and even outside such situations, dogs may experience a synchronicity of stress with human caregivers (Sundman et al. 2019). There are significant challenges for dogs—who are not able to consent to being pets—in a role that can require extreme levels of emotional control and suppression of species-appropriate instincts and behaviours (Haraway 2008; Pierce and Bekoff 2021). In addition, there are significant challenges for humans living with dogs, such as the emotional, financial, and often physical toll of caring for animals who are living with physical or mental illness, struggling behaviourally, or requiring end-of-life care (Belshaw, Asher and Dean 2019; Christiansen et al. 2013; Laing and Mayela 2018; Pierce 2012).

It has been argued that the lower status of pets within human households is also evidenced by the fact that while pets are grievable they are replaceable (Redmalm 2015, 32). This is insufficiently nuanced because the kinds of relationships between people and dogs differs considerably. The replaceability of dog companions who may be surrendered to shelters or taken to vets to be killed and subsequently replaced with a dog of a more suitable breed/age/temperament is a relationship of abusive power. However, to replace a much-loved and grieved dog companion with a new dog is not necessarily so. Neither does it indicate human exceptionalism. Rather, it suggests that relations with dogs are usually time bound simply by the significantly shorter life expectancy of a dog. In addition, as the literature on personal life makes plain, human relations are replaceable also, with the increased liquidity of contemporary personal life in the West. Studies have found people will make considerable sacrifices because of their attachment to animal companions, for example, by becoming or remaining homeless (Irvine 2013; Scanlon et al. 2021). Additionally, observational work has suggested that despite a context of human power, dogs can, and can be enabled to, express meaning in encounters with human carers enabling the adjustment of their treatment (Motamedi-Fraser 2019).

In many households, dogs and humans experience mutually beneficial relationships characterised by care, companionship and interdependence that operates intra-actively between the species, rather than in one direction, as will be discussed in chapter 6 (also Cudworth 2022). In situations where the needs and wants of both human and dog household members are accommodated, the human-dog relation might be understood to be 'intra-active: emergent, fluid, a becoming together' (Price 2017, 11). Relationships between humans and dogs that are based in care and co-creation of home life can allow for the relative flourishing of humans and animal companions in ways that muddy boundaries between human and animal, as we saw in chapter 3 (also Cudworth 2021).

The literature on animal companions as family tends to situate animals such as dogs living in homes with humans by stretching ideas about the human family to incorporate nonhumans. This often involves over-comparing relations with dogs and others to relations with humans in ways that suggest anthropomorphism or surrogacy. Alternatively, species distinction and inequality are taken to suggest that dogs remain outside of family because the persistence of difference and differentials of power is taken as evidence of the ambivalent status of animals. There is a need to retain a focus on power differentials while, at the same time, enabling the wider range of emotional bonds and relations that the sociology of personal life is open to. Rather than stretch the notion of family, or retain its humancentrism by excluding animal companions, might kin be more inclusive and also more productive in thinking about relations across the species barrier?

FROM FAMILY TO KIN

> I believe that all ethical relating, within or between species, is knit from the silk-strong thread of alertness to others-in-relation. (Haraway 2003, 50)

> Shakespeare's constant word play on kin and kind infected me by high school and has been in the genome of just about everything I have written. (Haraway, in Potts and Haraway 2010, 321)

As Haraway suggests, both kin and kind are bound by the same root. While the dictionary definition of kin overlaps with family, the terms are not synonymous. There is overlap in that the term kind may refer to a family, a species, a biologically defined class or set of plants or animals whose members have the same or common characteristics. Kindred is commonly held to simply mean in close relation. This emphasis on relationality is important, for:

> It's not necessarily to be biologically related but in some consequential way to belong in the same category with each other in such a way that has

> consequences. . . . Nobody can be kin to everything, but our kin networks can be full of attachment sites. I feel like the need for the care across generations is urgent, and it cannot be just a humanist affair. (Haraway in Paulson 2019, n.p.)

Haraway's suggestion is that kin enables affiliation and belonging which can be understood beyond human links and attachments of dependency and emotional connection. In her work on companion species, Haraway specifically links kin-making to relations with companion animals and claims to write as a member of 'a fictive kin group in training' (Haraway 2003, 40). This work, kindled through her relationship with Australian shepherd dog Ms. Cayenne Pepper, explores the forms of connectivity, intimacy and relationality which have emerged historically and constitute current practices. Intraspecies relations are an exploration of boundaries—'Whom and what do I touch when I touch my dog?', she asks (Haraway 2008, 3).

More recently, rather than examine relations with specific species in specific locales, her emphasis is on who we choose as kin, as partners, collaborators and allies. Haraway asserts that choosing kin is transgressive as kin 'is a wild category that all sorts of people do their best to domesticate' with the Western conception of family being a key line of the domestication of kin into the formal and narrow, legally demarcated and often genealogically bound 'godkin' (2016a, 2). Haraway challenges readers to make kin, with persons not bounded by ancestry or genealogy, outside of normative familial and species-bound structures, and to 'practice better care of kinds-as-assemblages (not species one at a time)' (Haraway 2015, 161). While Smart's sociology of personal life claims to reflect the bottom-up realities of varied chosen relationships, Haraway's making kin is more political—it is an ethical path for better species interrelating, where the strangeness of other species is accepted, and some of the closest of our kin are not human.

In the final chapters of *Staying with the Trouble*, Haraway (2016a) produces some speculative fiction. Her 'Camille Stories' tell of people living in a time of ongoing extinction due to climate change, the effects of which last for centuries. For her imagined 'Children of Compost', kin-making involves intentionally reducing the numbers of human babies reproduced and promoting the flourishing of species threatened with extinction. These ideas are developed in Haraway's manifesto call to 'make kin not babies' (Haraway 2018). Haraway (2017) admits that the 'not population' element of her 'making kin' slogan has got her 'in a lot of trouble'. There has been a concerted attack on her use of population, a concept which some consider irredeemably linked to imperialist framings, racialised violences and the 'economization of life' (Murphy 2017, 135); and others suggest it is anti-feminist in its 'dangerous neo-Malthusian logics' (Ojeda, Sasser and Lunstrum 2020, 327). While Haraway (2018) makes clear that it is different populations of humans who have varied patterns of resource use and environmental destruction, she has been accused of suggesting that some are to be more valued in the face of environmental threat than others (Dow and Lamoreau 2020, 481). She is also criticised for being adultist in presenting children as passive by 'being made' and childist by suggesting that having children is inevitably negative in terms of environmental consequences (Mattheis 2022; also, Hohti and Tammi 2019; Taylor 2020). My reading of Haraway, however, is that she is highly critical of the way children who are not of the 'right kind' are treated (2018, 96) and sees making kin as a project that is pro-child in that babies from all communities might be rare but valued. Opening the notion of kin through drawing on decolonial, Indigenous and queer perspectives in order to 'tell new stories to sustain needed new worlds' is important (Haraway 2018, 98; see, also, Benjamin 2018 and TallBear 2018). It is also notable that critics have little to say about the aggressive pro-natalist heteronormative familialism of nation state policies, and white settler cultures or the violent reproductive interventions against less 'desirable' populations that is a key thread across Clarke and Haraway's collection *Making Kin not Population*.

I think that Haraway's ideas about making kin can be extricated from the debate on population (although I think that she would adamantly oppose such a suggestion) and can be considered as a more inclusive alternative to the idea of family. However, I also am concerned that she is hugely ambitious in suggesting queer non-natalist kin-making as part of the solution to climate catastrophe and building more nourishing worlds.[2] Haraway acknowledges that we cannot be kin to everything, that not all relations are positive and we need to think critically about the kinds of relations we make or break. However, species difference needs to be writ large into the notion of making oddkin. There are varied forms and intensities of kinship possible across the species divide and a differentiated notion of kin is helpful. While Haraway calls us to relate to the microbial world, the quality of the relations is inevitably different (bonds of kin may be less emotionally proximate) than those with species which are less odd to us. For example, dogs and cats inhabit home homespace and share lives in ways more intimate and response-filled than those with other species.

In her earlier companion species work, Haraway's observations on everyday relations and practices are used to illustrate connection, affection and co-constituted practice in lives with other creatures, particularly dogs. In many ways, this is closer to the bottom up, grounded relationality that the sociology of personal life seeks to encompass. Charles contends that it is 'daily practices of kinship' in shared human-dog households which enables dogs to be constituted as 'kin' because they 'construct a world where dogs and humans are part of the same social group' (2016, 9.3). Others working in sociology have been attracted to Charles's approach to intraspecies kinship as opposed to Haraway's position, as the latter has been seen to emphasise hybridity (Ashall and Hobson-West 2017, 910); yet Haraway has for some time moved away from understanding relations between humans and other species in terms of companion species and has been drawn to kin and kin-making. Admittedly, Haraway understands kin far more broadly than does Charles. Charles uses kinship as a more extensive concept than family, but uses it very specifically as a way of describing deep connections between humans and animal companions in the home. That Charles allows dogs to be conceptualised as kin but, as we have seen above, not as family is significant because it illustrates the problem with narrow ways in which familial relations are cast sociologically. Haraway's understanding of kin draws on indigenous thinking, influenced by her engagement with the work of those such as Deborah Bird Rose, and is far broader encompassing relations with all kinds of other animals, peoples, plants and 'country'[3] (Haraway 2016a, 2022). Haraway also suggests kin might be broad in scope, going so far as to suggest that 'all earthlings are kin in the deepest sense' (2016b, 103) and that reinventing kin-making is key to responding to ecological crisis.

A New Zealand study of family relations with animal companions using storytelling methods found that people face daily challenges to make sense of what both they and their animal companions are doing in their specific situations and interactions. They also found an interesting difference between indigenous and non-indigenous participants when it came to relations with animal companions. Contributions from participants from indigenous communities involved storytelling where animal companions were referred to as 'furry whānau' (Sayers, Forrest and Pearson 2022, 14). The latter, is a broader category than family, encompassing various human relationships and other affinities. Sayers, Forrest and Pearson conclude that situated caring knowledge, love and reciprocal care relationships across species ultimately suggests that the conception of family 'needs revising', but their data suggests we might draw on other worldviews in so doing (2022, 14). Indigenous scholars have problematised conceptions of family as colonial impositions and, like Haraway, promoted kin as more inclusive (TallBear 2018). Haraway is ambitious in suggesting kin-making as the solution to climate catastrophe; rather, a range of radical responses will be required (see Cudworth and Hobden 2018). I also think a range of concepts are needed, too, and suspect the kin might not be able to do all the work for us. Most significant for the discussion in this chapter, however, is the issue of species difference, because there are different forms and intensities of relations possible. While Haraway calls us to relate to the world of compost, bonds with plants or microbes may be less emotionally proximate. If bonds are less proximate with other species, can this be called 'kin'? Here, sociological work on intimacy might usefully intervene. 'Loving, caring and sharing' are dimensions of intimacy suggests

Jamieson (1998, 8), and there are some species with whom such exchanges might more readily happen.

As much translation is required in making relations across species, familial metaphors may not best serve us in thinking about personal relations and emotional connections with animal companions. The following sections draw on the ethnography to consider how the situated, embodied experiences in everyday practice in human-dog households shed light on how the Western conception of family might be revised, and how kin might be a more productive category in understanding affective ties and relations within multispecies homes.

LOVE, FRIENDSHIP AND KIN-MAKING BEYOND THE HUMAN FAMILY

This section focuses on the stories people tell of their relationships in terms of how they understand the role of dog companions within their homes. Different degrees of intimacy and affectional ties are present, and there are varied levels of ambiguity and tension in the relationships. For some, there is much concern about dependency and lack of dog autonomy; for others, there is a special quality to the kinds of relationship we develop across the boundaries of species, which is odd, challenging and sometimes 'magical'. Other humans resist the idea of 'the dog' as a species, stressing the individuality of their relationships with different dogs as unique individuals; while some consider that 'the dog' is categorically significant, encouraging humans to think about species difference. In the quoted material below, I include sexuality of respondents where it may have an impact on what they say, or to prevent assumptions by the reader given the content.

Dogs and Non-Dogs—Connection and Relations

In thinking about dogs-as-kin, it is important to ask whether dogs occupy a different role to other animals which may be kept in the household as companions. Are some animals more likely to be kin than others? While dogs and cats are the preeminent animal companions, also popular in the UK are birds kept indoors (mainly parrots), guinea pigs, hamsters, tortoises and turtles, lizards, hens and snakes (Statista 2022e). Other reptiles, horses, donkeys, llamas, alpacas and pigs are also kept as pets; and fishes—kept both indoors and outdoors—are commonly referred to as pets (RSPCA 2023). In the UK, Indian stick insects are sometimes kept as pets and described as such (RSPCA n.d.). However, in comparing the stick insects kept by his young son and his dog, one man reflected that 'you can't have the same kind of relationship', as this interaction with his (female) co-interviewee suggests:

> M: Stick insects are a bit beyond me. But maybe people feel like that [the way he feels about his dog] about their goldfish as well. But it's more extreme with mammals, I guess. It's more evident; it's more clear.
>
> F: Like with stick insects and fish it's that closeness to a living thing, possibly, but not the relationship side. You know, the interest of watching what they're doing, the way they move, the colours, and the way that they breathe—it's the visual thing, the interest in nature not the relationship, I think.
>
> (M and F, friends, London)

A number of those interviewed kept fishes which they had inherited (with a property or from family) or had purchased. In his well-known work on the pet, Tuan (1984) reflects that fishes are considered ornaments, kept in the home or garden, and an interviewee living with fishes made a similar point:

> the fish are kind of part of the home, really, they're part of the furniture and most of the time I take them for granted. I mean, they're there . . . I'm so used to them, to be honest with you I don't, don't really have a relationship. I think it's because with the fish you can't physically be in their world. (F, London)

Here, the extent to which animals are considered odd or distanced from humans impacts the closeness of the relationship and thereby the extent to which they can be considered 'kin'. Interestingly, however, it was not only the species or the keeping of animals outside that necessarily affected the emotional connection with animals. While the woman above described the tropical fishes kept in a tank inside the home as like furniture, she took me outside to see Koi carp kept in a very large pond which took up much of the garden. Unlike the small fishes in the tank, the Koi were described as 'friendly' animals who 'come up and greet you and they like to eat from your hand'. Having said that people cannot be in the world of fishes with respect to the indoor animals, however, this woman described how in warmer weather she would sit by the pond with her feet in the water and that the Koi would rub against her and could be stroked. Koi are often described as fish that are particularly habituated and responsive to humans (Fife-Cooke and Franks 2021), and it was the ability to engage in individual relationships with the fishes, to touch them and to 'be in their world', which seemed to enable a different relationship with fishes of a particular kind.

More than a third of interviewees lived with both cats and dogs. Most differentiated the level of responsibility, with dogs requiring more time-consuming care, but close relations were always demonstrated with cats. Others had close relations with guinea pigs such that in one case, guinea pigs along with dogs, were exchanged by two sisters who undertook day-care for each other's animals when they were working. However, there were exceptions to close relations with mammals, with troubling accounts of estrangement from another animal popularly (yet often poorly) kept, rabbits, recalled by a number of interviewees from their childhood:

> Our rabbit was called Tiger and it was [pause] just awful; it was like a pit bull version of a rabbit. We were frightened to death of it; my sister was absolutely petrified of this rabbit. Umm, we lived next door to a church and the rabbit would frequently go into the churchyard and just sit in the middle of big graves waiting for us to try and catch her . . . she would bite you and scratch you and was very unpleasant. (F, London)

> We had a rabbit [when we were] growing up. It disappeared. My Mum said it went to a farm. I don't know, I don't think it did go to farm. I don't like to think about it. I don't think we treated that rabbit very well. (F, London)

The contrast between this and the strength of connectedness to dog companions is stark. Some interviewees suggested they would be prepared to go to extreme measures to care for their dogs, with a number mentioning that they would surrender their home to pay for medical costs for animals or, in this case, stay in another country:

> Do you remember [to partner] that couple who stayed in France because they couldn't get back in [to the UK] with their dog? The vet hadn't done the paperwork properly, so . . . they went back, and they've been in France ever since! . . . If we went to Spain and the dogs couldn't come back, then we wouldn't either. And they [people without dogs] think, 'It's a dog!'. People who don't like dogs or don't want dogs don't understand at all. (Lesbian couple, London)

People mentioned grieving over past dogs both as adults and as children, the grief their parents experienced over the loss of dogs, their fear in relation to the illness and death of their current dog(s) and the extent to which dogs had been an integral part of the fabric of their lives over an often very extended period:

> I always had dogs since I was two. [My parents] bought me a dog because I was the middle child and they thought I needed something special. A special relationship [laughing], so I had always dogs, yes. And then gaps because I was so [pause] so distressed after they'd died. (F, London)

They were also very clear that what was involved was a very intense relationship, of an intensity that was often compared to bonds with human kin:

> What do you do about that? What do you do about any person? Do you stop loving so much and put the barriers up? No, it's going to happen [the death of her dog] and you've just got to go with the flow . . . I'm just sure that the joy outweighs the grief. I've just lost my dad, I've lost my uncle, you know, I know what it means. (F, London)

In understanding multispecies kinship, species matters, as does particular kinds (such as the difference between Koi and other fishes). Different proximity and sharing of spaces affect different levels of connectedness and intimacy. One interviewee had kept a range of species as companions, including hand-reared sheep that had access to her home and played with her and her husband, and a litter of hand-reared rabbits that were fed with

> a doll's bottle, and they used to come everywhere with me, because I'd got to keep topping them up [with milk], and you know, you'd come out [of the shops] and all these rabbits [would be] sitting along the back shelf of the car [laughing].[4] (F, Leicestershire)

While this woman considered a wide range of creatures could become close companions and share homespace, she was very clear that understanding the needs of particular kinds of animals and treating them appropriately was 'the priority'. This suggests that making oddkin is implicated in understanding difference and specificity—of both possible connectedness between species, and the difference of specific kinds of relationships with individual animals. Making multispecies kinship means resisting the flattening out of relations with nonhumans.

Dogs, Parenting and Family Values

While there are comparisons to be made with parenting, the human participants in this study showed strong resistance to understanding the relationship with dog companions as a form of parental guardianship, and aversion to the infantilising of dogs, who were seen as agentic members of the household 'with a point of view'.

As family is a persistent cultural concept, and in heteronormative, biogenetic Western white culture is often conflated with the household, it is not surprising, however, that those with close relations with the dogs they live with extend the concept of family to include them. Family was an element of the narrative in the field notes, and in living with and knowing dogs who are biologically related, family tensions were in evidence:

> Next stop is Gypsy's brother, Percy, who ignores her as usual. . . . We talk about who likes who. Gypsy loves everyone, Percy likes Opus and hates Flash. I mention nephew Syd. 'Percy hates him!', she exclaims. She thinks it's because Flash is Syd's dad. (Field notes)

As I write this, a framed picture of dogkin is to my right: Gypsy, with older brother Opus, her nephews Rondo and Syd and her father Toby; while pictures of Gypsy and favourite brother, Tlaquito, and niece Calita are in other rooms. It raises an interesting question of relationality and human power, for most dogs do not get to maintain the relations they have with parents and siblings. Rather, they are expected to become part of human households where they are the only dog, or are most usually unrelated to other dogs in the household.

A number of interviewees described relations with dogs as familial:

> They're just part of the family, aren't they? (F, Leicestershire)

> both my husband and myself grew up with dogs, always part of the family, family's not complete without a dog. (F, London)

However, many struggled to describe the precise kind of relationship. Here, a woman moves through a number of conceptions before moving away from family:

> I'm not treating them like a person. . . . Because I don't think of them as children, I've never wanted children. But they're part of family [pause] or *life* and they have a place, and they have to be treated in a certain way. (F, lesbian, London)

Others use the similarity between relations of care and affective connection between people and children to try and articulate their relationship with their dog:

> The love people say that they feel for their children, I couldn't really understand it, but, you know, I think I might understand now a little bit. Because, you know, really, well, I just love him [her dog], I love him so much. It's incredible. (F, heterosexual, London)

> I have to put up with their [friends'] children and they have to put up with my, erm, appendage, erm, friend. (F, heterosexual, London)

In the literature focusing on dogs as family, and specifically as 'dog parents' in the US context, the cultural frames of teaching, training and caring for human children are used by some to construct their relationship with animals as parenthood (Owens and Grauerholz 2019, 11). The idea of dogs as children, however, was rare, only two interviewees claimed they saw their relationship as similar to mothering, or that their dog fulfilled a parenting need once a child was older. However, in both of these interviews, the overwhelming thrust of the narrative was that dogs should not be treated like humans or like children; rather, dogs are:

> an important member of the family, but they have to have their freedom . . . let them be a dog first, they're not there [for people] to have someone to control . . . I don't try and make them into a human being, silly hats at Christmas and stuff like that, I'm not into any of that. I don't control them, I let them be as free as they possibly can. (F, Leicestershire)

The most common way for notions of family to play out was by being attached to a story seen as humorous, or presented as a joke and accompanied by laughter:

> a surgeon at work, who wanted us to stay late to finish a case in theatre, and I said, 'Well, you really need to tell me how long you're going to be and how long you want me to stay because I need to get home for the boys, I can't leave them on their own all day', and his face, he was just like, 'What? You've left your children at home on their own?' [laughing]. (F, Leicestershire)

> he was the runt of the litter, and he had the same hair colour as me so [name of son] thought that we should get him [the dog] because he would think that I was his mother [laughing hard], which he does, obviously! (F, London)

There was a significant antipathy to the humanising of dogs in terms of practices which interviewees regarded as inappropriate or cruel. This was apparent in discussions of dog training and socialising, where enabling dogs to 'retain their dogginess and sense of adventure' was very important and not having a dog be 'squashed' or 'crushed' by excessive human regulation or 'rules' was desirable. It is important to point out that this was not representative of a cavalier attitude with respect to training, as socialisation with other dogs, other animals and humans, and training to ensure the safety of dogs and of people was understood as a necessity. The dressing up of dogs was singled out for particular critique:

> I've got a friend who's got a little Shih Tzu and we often walk together, and I have to say I think she's now treating her dog like it was a human, she's bringing it out in clothes, you know, Dorothy's [the dog] got more clothes than I have, erm, she actually treats the dog like she's a little girl. In fact, sometimes she ignores the baby and makes a fuss of the dog so the dog doesn't get jealous.

> I think it is important that you that you do remember that your dog is a dog. (F, London)

While this can be seen as inscribing the species boundary in the way Charles and others suggests happens, I think that it is also part of a narrative of the appreciation of the difference of species being linked to respect for dogs.

There was a gendering of the hostility to the notion of dogs being seen as surrogate children by those who were childless, in ways very different to the findings of work undertaken in the United States. The hostility was presented in ways that segue with recent and established feminist critiques of pro-natalist heteronormative family:

> Oh, we know what people think of us, the way they look at us, two women with a dog that's spoilt. I think dykes get that a lot, you know, 'It's their child'. It makes me furious. (F, lesbian, London)

> I think people always want to draw a parallel, and when you're my age and lots of your friends have had children, and your family are thinking that you should be having children—you're that age—then I think it's nice for them to put you in that box and say, 'Oh [interviewees name] hasn't had children yet, but that Rocky [name of dog] is her little baby'. I think almost it's just for them to reinforce their family values or something. (F, heterosexual, London)

All interviewees discussed the differences between relationships with dogs and with humans, and that what was special or particularly enjoyable about relations with dogs was precisely that dogs are not human. People often found this difficult to articulate:

> it's a different form of relationship that's so enjoyable because, I don't know, it's difficult to describe, isn't it? Erm, there's so much joy that can be through, you know, interacting with your dog because it brings out just a different side of your personality. Obviously, you don't have, um, the same language to communicate with them and so your relationship is different, based upon that lack of understanding, I suppose. You have to try and work out what each other are thinking. (F, heterosexual, London)

A strong narrative across the interview data is that what makes the specific relation between dogs and humans living together, is that they are attempting to negotiate everyday life across species difference. It is the cross-species interaction which leads some people to resist the comparison with children or other human family:

> I don't like that phrase that a dog is like a baby or whatever, because it's not . . . you've got this relationship with something that's not human. Yes, it's that thing with a different being. (F, heterosexual, London)

These kinds of cross-species connections and bonds, and also the differences and difficulties of intraspecies relations were important elements in my human participants' accounts of living with dogs. Dogs, and in some cases a range of other animals (cats, chickens, goats, guinea pigs, horses, sheep), were seen to come into a relational field where 'intelligence', 'emotions', 'care', 'play' and 'kindness' were aspects of shared lives, alongside different kinds of tension and misunderstandings. The success of these relationships demands much of dog companions, in a precarious position where reading their humans well is often key to their security and sometimes survival.

Most human participants reflected on the challenges of relating across the species and their attempts to better their understanding of dog behaviours, and of their particular dog. Some recent material in the dog literature suggests that human dog owners underestimate dog abilities and potential for independence (Pierce and Bekoff 2021), but my participants were more likely to reflect on the ingenuity and intelligence of dogs. For example:

> I have a high regard for dogs. I think their understanding, their comprehension of language—they understand us and our language—it's far superior of our understanding of their language. I think we completely underestimate their understanding of us. (F, Leicestershire)

> They're more intelligent than people think; they can make it very clear what they want. (F, London)

> I think he's very intelligent, he knows exactly what he's doing [dog barks]. (F, Leicestershire)

Like other research in the UK context, the use of parenting metaphors applied to relations with dogs, along with other familial metaphors, is ambiguous (Charles 2016, 5.4). In addition, there was significant resistance to the infantilising of dogs and to 'family values' with respect to how relations, particularly between women and dogs, are culturally perceived. There was also some resistance to the idea that people were in relations with dogs collectively, as a species—rather, the relationship described was always with particular individual animals. At the same time, and as we will see below, there was also quite some reflection on what it means to be in relation across the species.

A Special Relationship

Irvine argues that despite the dependence of animal companions on human guardians, dependency 'doesn't necessarily translate into domination' (2004, 28). Certainly, domination is not how I think my human participants would understand their relationships. Interviewees were asked to describe what their dogs were like, and often gave detailed accounts of the specific personality of their dog, their likes and dislikes, their responses to experiences and relationships, such as:

> to smell foxes, chase squirrels, rabbits, that kind of thing . . . that's what makes her happy, so I think she's very, very much an outdoor dog, and she loves swimming. (F, London)

> She loves aftershave and she's very urban. She hates getting her feet wet. She doesn't see her role as a protector. She's a receiver, she's not a giver. (F, London)

> I can imagine Lola in a previous life being, being the sort of, erm, er [pause], you know, a bar room lush[5], with too much make up on and a dress that's a few sizes too small, you know, going for every fella in the place. She's only got to see a builder or a bar keeper and she's all over him like a rash [laughing]. (M, London)

Such accounts are anthropomorphised, yet as animal studies in the humanities and social sciences has long pointed out, anthropomorphism is a condition of our relations between humans and inevitably a component of our interactions with nonhumans. As with the findings of other UK studies, dogs were understood to possess selfhood and to be individuals with distinct personalities and characters (Charles 2016, 7.2). They were understood as active agents able to make their views and experiences known and, if needed, to articulate displeasure. When two women recount the experience of their dog being given a buster collar by the vet to prevent her licking a wound, they do so to illustrate the dog's sensitivity and emotional response:

> She got in the car with this sort of lampshade [buster] collar, and it was high drama, wasn't it [to partner]? And in the end—

> [Partner interrupts] I'd have had to have had a bloody Valium[6] if she was going to keep that thing on her head because she was just really distressed. She had a sore belly so we put cream on it, and we took it [the collar] off but she wouldn't talk to us for three days. We'd walk into a room, and she'd walk out. (Lesbian couple, London)

All interviewees considered that their dogs were in communication with them, and that they were able to understand at least some of what was being communicated vocally, facially and in terms of other readings of dog body language (as Motamedi-Fraser 2019 suggests). Relations with dogs were clearly relationships with individuals, and people who had experience of living with a number of dogs across the life course and/or, at the same time, were careful to specify particular personalities and bonds.

Most people also referred to dogs as creatures that they loved, or that they were in love with. In speaking of her relationship with Cayenne, Haraway reflects that she had a 'love affair with her' (in Paulson 2019, n.p). Marjorie Garber suggests that the close emotional ties people have with companion animals such as dogs should be understood as love, as close relations mean people need to make a leap of the imagination in trying to understand the emotional lives of another species. Garber is clear that relationships with animal companions are not substitutes or rehearsals for human relationships, but they provide a different and specific relationship (1997, 32). In living with other animals as companions, we are forced to understand the inner being of another species. In this process, we access the minds of the animals that we live with, and this accessing process is inevitable, partial and contestable.

There are different ways in which dog love is expressed, but the word 'love' was used often by participants in this study, and repeatedly.

> We just desperately love them, just *love* them. (F, London)

> You just love their [pause] their *being*. (F, London)

In some cases, people had a preference for a particular breed of dog, and some described themselves as 'in love' with lurchers, spaniels, 'rescue mutts with collie in them' or, in the case of this woman:

> I always loved Old English sheep dogs, I just loved the way they walk, how the hips all wobble and they're silly and the way they sit with their leg out. I just like that breed. I love the breed. And they've got a lovely characteristic. . . . I just love them, love them. (F, Leicestershire)

However, the story of a love for a particular type of dog quickly turns into a story of a history with different dogs, the first of which 'was a gentle giant really, oh God, I think I'm going to cry'. This was followed by a reflection on the devastation of losing a specific dog and the challenges of grieving and establishing relations with other dogs, over years. This pattern of discussing the kind of dog followed by the specific dog was common. For a woman with a long history of living with multiple rescue dogs of various breeds:

> I love the Staffies.[7] The one that came out of a bag from the canal [where he had been abandoned as a puppy], I loved him to bits, he was seventeen when he died. Yes, I've loved them all, they've all been lovely in different strange ways. (F, Leicestershire)

People would comment on who a particular dog would be closest to in their household, the specific and different relationships dogs had with other dogs and other humans they knew, or the different qualities of their relationship with a particular animal. In some cases, the way in which the latter was discussed was reminiscent of the ways in which people might discuss a lover:

> It's really hard to put into words, because I just have so many feelings. When I'm just thinking about something, thinking about him during the day when he's not there, something that's made you smile, you know, he just makes you laugh—it's all quite overwhelming, really. (F, London)

> It wasn't love at first sight to be honest, but it is now; I sit here in the evenings looking at her, thinking she's very beautiful. (F, London)

However, just as people made distinctions between dogs and children, they were clear that relationships of close emotional attachment to dogs was specific because it involved the challenge of species difference:

> I think there's something really magical about becoming intimate with a member of a different species. And the bond of, you know, need, and dependency and responsibility and love crossing that boundary, is very exciting. In some ways, even more exciting than with another human, 'cause it's so much more mysterious as well. Isn't it Darling [to his dog]? It's mysterious! (M, London)

> It's so hard, to get your head around it . . . that she's a dog and that I am [pause] I am a human because she is so h- you know what I mean, she is so human in so many ways. I just look at her with real love and I just think god—you know, 'You know me and I know you and you're another, you are another [pause] that thing, you are another species'. Species, it's a very strange thing. (F, London)

People understood the relationships they had with dogs to be reciprocal—in terms of enjoying each other's company, or shared activities such as walking. As we will see in the following chapter, they also understood elements of caring relations to be reciprocated. People also considered that in terms of affective relations there is 'a real reciprocity there'; they felt their love was returned. While some did suggest that the love of a dog was 'unconditional', this was not commonly used by interviewees. Relationships with particular dogs changed over time and were recounted as developing in particular situations. Some relationships were not straightforward but were hard work—particularly with nervous dogs and those experiencing trauma prior to rescue and rehoming.

The greatest difficulty in having emotional bonds with dogs, however, was often described as 'the end', with one interviewee abruptly ending a discussion, saying, 'Oh god, yes, the end. You can't ask about that', while turning and walking away. This quote is from my first interview and while I was careful to avoid discussion of death and dying, if interviewees came close to discussing these issues, they would become visibly emotional. It is a slight to suggest that dogs are not family as they are replaceable. Rather, their illnesses and deaths cause trauma and grief in ways some experience similarly to the loss of human kin. In the course of walking with dogs and meeting and talking to other packs of walkers, I know the death of a dog to be highly charged and emotionally impactful, not only for those who have lost their dog but communities of dogs and people who walk together, and also in relation with a particular animal. While 'the end', or coming close to discussion of it, prompted emotional reactions, the interviews as a whole demonstrated strong levels of emotional bonds between people and the dogs who share their homes. Returning to where we began, and refugee humans and the animals-of-their-heart fleeing or enduring war together, the reason why being able to take or having to leave animal companions brings joy and heartbreak is because of love. When a young refugee from Damascus was interviewed by the United Nation's refugee agency (UNHCR) in 2015, having arrived on the Greek island of Lesbos after walking more than three hundred miles with his puppy, who he carried for most of the journey, he simply said: 'I love this dog, I need her'. (Dearen 2015)

DOGS AS (ODD)KIN

> let's face it, you're all a bit odd aren't you, my furry little friend? Odd kin. (Koontz 1988, 12)

Adele Clarke suggests that the project of making kin involves reproduction, making new relationships and 'the daily actions that transform partial relations into deeper ones' (2018, 32–33). Relations with dogs are certainly those which transform over time through daily interaction. Power (2008) argues for a broadening of the concept of family beyond biological relations to include more-than-human relationships forged through cohabitation and everyday interaction in shared space. However, kin might be more inclusive and better capture the qualities of relationships in multispecies households than the concept of family. The otherness of dogs was a key narrative in my ethnographic material—they were not furry humans for my participants, rather, the *dogginess* of dogs was accommodated. People and dogs are engaged in relations with strange others—kin-making across species divisions. In light of this, I would suggest that the concept of the family is too humancentric and burdened with genealogic framing. As cross-species relations are necessarily fraught with accommodating very different ways of sensing and understanding the world, I also think oddkin is a concept that is a good fit for dog–human kin-making.

Oddkin can be useful to capture the close relationships some people and some dogs develop first, because relationality is linked to 'response-ability' (capacity to respond) and to reciprocity in Haraway's formulation (2018, 70). Certainly, those human participants in this study saw care, affection and friendship as responses to the dogs they lived with and understood these as reciprocated in certain ways (as will be discussed in chapter 6). Second, because stretching the imagination is central to the process of making dog–human oddkin in ways human participants in this study described as interesting, strange, and magical, because 'kin are unfamiliar' (Haraway 2018, 94). Haraway's model of kin-making draws on indigenous thinking and concepts such as the Australian aboriginal conception of country as a multispecies network and co-constituted combination of beings in place and space. This is a third reason why oddkin-making is a good fit for human–dog relationality, because as we saw in chapter 2, this project found intraspecies networks of care and response-ability, which constitute particular posthuman communities of people and dogs who walk together in the same space/place and demonstrate concern for other species and communities which share these spaces (see also Cudworth 2017).

While a minority of interviewees used the word 'family' to describe relationships with dogs, none used metaphors identifying themselves as parents or dogs as children except in jest. Unlike findings from some US research (Laurent-Simpson 2021; Volsche 2018) there was generally antipathy to the idea that people were in a parenting role with animals, and it was important to enable dogs to be free to express themselves and 'be a dog'. I remain unconvinced that treating animals as children means that they are such, or that acting towards another animal in a parenting role necessarily broadens family membership. Rather, the humanising of dogs (and others) has welfare implications (Pierce and Bekoff 2021). The concept of families of choice still ties us to the overlaps with blood relations (Weston 1991), and this precludes other animals, despite opening up kin-making as an active process. The otherness of dogs was a key narrative in the stories of human participants—dogs were not furry humans, rather, the dogginess of dogs was accommodated, and people demonstrated high regard for dogs—what Blouin (2013), whom we encountered in earlier chapters, would refer to as 'protectionistic' attitudes.

People and dogs are engaged in relations with strange others, kin-making across species divisions.

This said, I do have some reservations about the flexibility of the concept of kin. While feminism needs to revisit, develop and promote different visions of kin-making 'across species, generations, across categories' (Haraway 2017), I do not think understanding dogs as kin for some humans will necessarily benefit a planet in crisis. However, I am glad to see that multispecies kin-making for Haraway needs to be in tandem with an end to the forced reproduction of animals raised for food in industrial agricultural systems (Haraway 2018, 73). Changing the lucrative part of the animal-industrial complex that is the pet food industry might certainly alleviate some of the significant environmental burden of dog companions. Finally, and perhaps ultimately, I think oddkin-making is a product of Haraway's passion for dogs specifically, and one dog in particular, Cayenne, who Haraway (2017, 36:53) describes as 'a dog I fell in love with and who completely changed my life'. While as Haraway suggests, 'kin is a wild category which all sorts of people do their best to domesticate' (2016a, 2), what happens in the love affairs between some dogs and their humans is a co-constituted process of domestication and re-wilding. As we saw in chapter 3, people blur boundaries in shared home lives with dogs and 'become more dog' in their domestic arrangements. Making multispecies kinship means resisting the flattening out of relations with other creatures and understanding human-dog relationality as a co-constitutive process which has specific qualities, despite taking place in the social context of human domination. In the stories people shared of their relationships with dogs (and sometimes other animals) what was crucial was a particular relation with specific animals. In their everyday practices, relationships are developed over time so that people are able to give detailed accounts of the personality traits of the dogs they share their homes with and the kinds of relations they have with them. These relationships are intense and involve a range of strong emotions concerning care, companionship, happiness and loss that sociologists of intimacy would recognise (Smart 2007; Jamieson 1998) and which question the idea of individual dogs and relationships with them as replaceable in ways very different to human relationships (contra Redmalm 2015). They are seen as reciprocal with dogs acting to confirm, to change and impact the quality of the relations with their humans. Inevitably, people anthropomorphise the animals they live with in describing characteristics, relationships and behaviours.

It needs to be recognised, however, that in human homes and subject to human rules and schedules, most dog companions do not have the opportunity to be a dog first. Open-ended conceptions of personal relations tend to elide an understanding of how dominatory power is reproduced, yet multispecies kin-making reflects a human-dominant world even in the most power-attentive cases. This said, I also consider that it offers more possibility for change than shoe-horning animal companions into the 'family'. Thinking about dogs as beings with views about the world and a perspective that is shaped by both their species and their social context might guard against their infantilisation, inappropriate treatment and backlash when dogs behave more like a dog and less like a (humanised or human-deferential) companion. While a minority of interviewees used the word 'family' to describe relationships with dogs living in their household, none used metaphors identifying themselves as parents or dogs as children (or other roles for human family members), except in jest. In fact, this study found there was quite significant hostility to the humanising of dogs in this way. Perhaps it is because of the use of dogs (and other creatures) as pets—which usually involves a (more or less) violent process of removing them from parents and siblings when very young—that people extend family, as a metaphor, when discussing their dogs.

In seeking to shift sociology away from family to a broader notion of kin, there is admittedly tension between the extent of the conceptual move and the evidence from a qualitative study such as this. Such tension attends the sociology of the family and personal life more generally with a gap between 'broad, generalised theoretical statements and small-scale, detailed empirical study' (Smart 2007, 8). Participants understood dog companions to be close relations, involving the blurring of the human-dog boundary and experiences of love across the barrier of species. This does not mean that the species barrier is overcome, far from it. However, it does suggest that despite a human supremacist public social world, private spaces of home and the intimate relations therein might reflect different kinds of practices and relations. In addition, participants did continue kin in ways that were challenging for humancentrism and were resistant to the notion that dogkin were substitutes for human relations.

I do think that family is hard to uncouple from genealogy, heteronormativity, natalism and humancentrism. It is not surprising that some of the research on multispecies households seems more comfortable with concepts of kin than family (Charles 2016). Studies conceptualising dogs as family do tend to be characterised by narratives of infantilism, humancentrism and, sometimes, heteronormativity (Greenbaume 2004; Volsche and Gray 2016). Making kin is, to my mind, a better descriptor for what goes on in forging relational ties with dog companions, but it may work better for dogs and other mammals than for other creatures—some species were too odd to be kin for some human participants in this research. This leads me to conclude that while I have much sympathy for Haraway's notion of kin-making, we need to be specific and careful, to tease out different elements of relationality in our response to various very different species. Investigating how oddkin are made and how relations are sustained and broadened will be fascinating work, but it will need to recognise that species difference needs to be paid attention.

Chapter 6

Working and Caring

This final chapter focuses on the experience of people living with dog companions through the nexus of work, labour and care. Nonhuman animals have rarely been understood to engage in work, but some human-animal studies scholarship has recently turned its attention to animal work. This chapter explores the possibility that animal work may enable new insights into the role of companion dogs and human relations with them. It also, however, assesses whether specific kinds of work might have most potential for enabling improvements in dog and human lives. The kinds of activities which concern this chapter are some of those familiar to feminism, such as balancing the demands of home and work and the forms and impacts of reproductive, domestic or caring labour. In addition, it asks questions of the kinds of work dogs specifically might be seen to undertake in human occupations.

Certainly, the subject of human work has been considered sociologically significant, and the public world of work has been a key area of sociological interest since the discipline emerged. Foundational figures such as Marx, Weber and Durkheim were preoccupied with work as formally recognised through capitalist relations (Lukes 1973; Ritzer 1992); while William E. B. Du Bois was concerned with both paid and unpaid labour (Zamir 2008). Claus Offe (1985) went as far as to suggest that work is the 'key sociological category' (1985, 129). The impact of degrading working conditions and the alienation of workers in Marx has been profoundly influential and developed by feminist scholarship to include emotional labour in the workplace (Hochschild 2012) and domestic or reproductive labour in the home (Oakley 1974; Weeks 2011), in the homes of others (Anderson 2000) and linked in global networks (Ehrenreich and Hochschild 2002). This enabled the caring labour undertaken for kin and loved ones—what I call labours of love—to be understood as work in addition to being both gendered and a relational responsibility (of mothering, for example).

Reproductive labour necessitated that the sociological lens includes everyday practices, but although these have come to incorporate a wide range of new subjects, nonhuman animals have been largely ignored. While carework has usually been understood to involve humans as both the providers and recipients of care; ecofeminism/feminist animal studies scholarship has demonstrated the material and theoretical significance of care, by humans, for other creatures and the natural environment (Donovan and Adams 2007; Cudworth 2022; Puig de la Bellacasa 2017). This chapter will suggest that carework is not undertaken exclusively by humans for other animals but, rather, that such work can be reciprocal and multispecies, in addition to having gendered characteristics. Some research has examined the conflicting demands of paid work and the family (Drobnič 2011), particularly carework for children and the tensions for some women around cultural notions of good parenting and the 'good mother' (Henderson et al. 2016). In light of this, a further question for consideration in this chapter will be whether such tensions exist in terms of carework undertaken for dogs, and in relation to cultural constructions of the 'good' or 'responsible' dog owner.

Ultimately, this chapter will argue that while work, broadly defined or linked to human occupations, has limited traction in understanding human relations with animals, carework has greater promise. The contribution of this chapter to debates on animal work is threefold. First, it draws on the ethnography to specify the kinds of everyday tasks human carers undertake for dogs, and how they negotiate paid employment in relation to this carework. This empirical grounding is absent in existing literature on animal work, feminist and otherwise. Second, the chapter suggests that carework in human-dog relationships is reciprocal or intraspecies, examining the ways in which we might understand dogs to be undertaking carework in companion species relations. This is hinted at in some feminist dog writing (specifically, Haraway 2008) but has not been elaborated or theorised. My ethnographic research indicates, however, that the carework of both humans and dogs can be understood as a set of emergent labour practices. In light of this, this chapter proposes that the range of reproductive labour tasks already familiar to feminism can both be extended to incorporate care for other animals and also to conceptualise some of the ways dog companions care for the humans with whom they are in relation.

Some animal studies scholarship has been attentive to animal work; including work undertaken *with* animals (such as by veterinarians, Arluke and Sanders 1996), performed *by* animals (Porcher 2011) and undertaken *for* animals (such as advocacy, Hamilton and Taylor 2013). Kendra Coulter argues for an inclusive conception of animal work, involving paid, unpaid and subsistence work (2016a, 18–19). These descriptive categories are useful for thinking about animal work in the first two sections of this chapter, which position the idea of dogs as workers and as careworkers within the emerging literature on animal work and the more established scholarship on carework.

Thereafter, the chapter draws on the ethnography in considering the ways in which human working practices affect the lives of dog companions and how human carers negotiate working life in relation to their dogs. It examines the carework undertaken by human owners for dogs, and the carework undertaken by companion dogs in relations with their humans. In framing the voluntary work of care, feminist conceptions of the material practices of carework will be key. While animal work may enable new insights into the role of companion dogs, however, the chapter also reflects on whether work is the most fruitful of sociological concepts for understanding our relations with other animals and considers the broader challenge of multispecies carework for feminism.

DO ANIMALS WORK?

Richard Scarry's *What Do People Do All Day?* first published in 1968, has proved to be one of the most popular illustrated books for children, and being a child of 1966, it was my favourite. The title appears a misnomer, given that there are no people, only animals, undertaking a wide range of human jobs and roles in 'Busytown'. In his meticulous content analysis of Scarry's text, sociologist John Levi Martin identifies twenty kinds of animal and 132 occupational categories, mapping the status of jobs to the sociologics of species (2000, 206). For example, feminised labour is overwhelmingly performed by feminised animals such as cats and rabbits, dirty and unskilled jobs by pigs (as exemplifiers of the working-class body), and authoritative jobs are undertaken by predators such as lions, bears and foxes (2000, 215–23).[1] What do dogs do all day? Busytown dogs are placed in trusted positions in a range of public services, particularly law enforcement. Levi Martin's well-made point is that there is a hidden language or code inscribed in children's books, which teaches children 'what kinds of people do what kinds of things' (2000, 225). However, the hierarchical categorisation of animals also speaks to ways in which real creatures are socially situated in naturalised categories. Scarry depicts pigs as greedy, dirty and unskilled, a comfortable popular stereotype, perhaps, given the particularly tortuous treatment of these animals in agricultural systems. The role of dogs in law enforcement is one we recognise beyond children's fiction, one in which these relatively privileged animals, commonly in closer relationships with humans, are seen to work collaboratively (Smith et al. 2021). This section is concerned with the possibility that animals can be seen to be working, what kinds of work they might do and how this can be conceptualised. Haraway has suggested that 'plumbing the category of labor' might enable positive improvements in the lives of animals (2008, 73), and the emerging literature on animal work has responded to this challenge. This part of the chapter also considers whether work is a useful category when applied to domesticate animals in general, and dogs in particular; and whether some kinds of work, specifically the work of caring for others, or carework, might have particular traction in understanding human-dog relations.

In thinking about whether animals can be said to work, some have considered wild animals to be working for their own species (Ingold 2012; Coulter 2016a), but most scholarship has focused on domesticate species used by humans in working contexts: for transport, hunting, guarding, security, assistance and food. When domesticated animals are discussed in relation to work, this tends to be in terms of human workplaces where people encounter animals as part of their work, such as veterinary surgeries (Clark and Knights 2018; Hamilton and Taylor 2013) or slaughterhouses (Hamilton and McCabe 2016; McCabe and Hamilton 2015). When animals kept as companions are discussed in relation to work, this has sometimes been related to public presentation, of celebrity[2] (Redmalm 2014) or political family (Skoglund and Redmalm 2016). Apart from assistance animals, dogs are generally excluded from, and understood to be out of place in, workspaces such as schools or universities and offices (Charles and Wolkowitz 2019). For Kendra Coulter (2016a), this is a narrow conception of animal work that does not appreciate the range of work activities with which nonhuman animals engage. She, like Haraway, has argued that thinking about animals as workers might enable positive improvements in the lives of animals.

How might such work be conceptualised? In relation to animal work, Haraway concludes that 'Marxist humanism is no more help in thinking about this . . . than other kinds of humanist formulae' (2008, 73). This is because for Marx, humans are unique in being able to work—to transform nature in ways that meet their needs, while labourers' bodies are not the 'nature' that is transformed (Dickens 2004, 258). Yet humans do work on and with their own bodies, contra Marx (Clark 2014, 152–55). Marx allows that animals undertake subsistence labour such as raising offspring, which he considers instinctual; but this has serious implications given that reproductive labour in humans remains, still, often unrecognised (Coulter 2016a, 64–65). I would not discount the significant influence of Marxism in the development of critical animal scholarship, however, and depart from Haraway here. Critical scholarship has usefully drawn on Marxist concepts in critiquing the violence and exploitation of farmed animals and, combined with postcolonial and feminist influences, enabling an intersectional politics with potentially emancipatory outcomes. Those working in animal studies have engaged with Marxist categories of labour, exploitation and alienation. Barbara Noske's concept of the animal-industrial complex, which we encountered in chapter 4, considers animals in farming contexts to be exploited similarly to women undertaking reproductive labour, having no boundaries to a working day. Farmed animals are alienated from the products of their labour (their offspring), their productive activity (being unable to use their full capabilities), their fellow creatures (through overcrowding or separation) and from surrounding nature (in highly artificial environments in industrial systems) (1997, 18–21). The animal-industrial complex captures networks of national and international agribusiness, linked exploitations of human and animal workers and decimation of the environment (1997, 24–31). In other accounts, farmed animals are said to labour 'under the same capitalist system as humans' (Hribal 2003, 436) and to be 'super-exploited living commodities' (Torres 2007, 8). In earlier work I have argued that the kinds of relations humans have with the animals farmed for food or bred for laboratory purposes should be understood in such a way, as oppressive and exploitative and devoid of potential in securing improvements in such animals' lives (Cudworth 2011a). However, it must be conceded that the application of Marx's concepts seems to have resulted in the loss of any notion of animal agency and of the use of concepts of reproductive labour outside sites of production, such as the farm.

Others have approached the question of animal work as cooperative endeavour involving animals as active agents. French farmer and sociologist Jocelyn Porcher considers the whole processes of domestication to be about 'inserting animals into human society through work' (2014, 1). In her manifesto for traditional 'husbandry', Porcher (2011) makes a searing critique of industrial livestock production, which normalises violence and commodities living creatures. Yet to abandon meat production is to seek a rupture between ourselves and other animals, in her view. Porcher's arguments rely on a romanticised picture of traditional farming and invoke reactionary sentiments (Stănescue 2013). She insists that traditional farming is predicated on a 'good life' for animals but does not have the answer to the question of the place of death as farmed animals' 'ultimate end of worth' (2014, 8). Until she does, it is unsurprising that some of us 'question as to why the whole system is not deemed animal abuse' (Coulter 2016a, 49). I concur with Nicolas Delon (2020) that Porcher fails to see any alternative to exploitation (through farming animals) or extinction (through its abolition), and that in making her argument for farmed animals as workers, she distorts the harm of animal death and greatly overestimates the value of animal work for the animals concerned. One point of interest in Porcher's contribution, however, is the notion of collaboration as work. She understands animals in traditional farming contexts to be co-operating with humans and to be co-producers in the food system. Her observational research on a dairy farm undertaken with Tiphaine Schmitt concluded that cows were working through modifying their behaviour to obey (or 'co-operate' with) certain rules set by humans and by investing their 'intelligence and affects in the activity of work' (Porcher and

Schmitt 2012, 56). Such labour is invisible and only becomes apparent on the occasions that cows are resistant or disruptive. Haraway has written about laboratory animals similarly as collaborative co-workers (Haraway 2008, 73), but as with Porcher's cows, such representation is disingenuous. Animals may co-operate but are perhaps more accurately conceptualised as 'worked on' rather than worked with (Weisberg 2009, 36).

Away from farms and laboratories, other animals might be seen as undertaking labour that better lends itself to the concept of work. Dogs are used, for example, for their sense of smell in finding and searching at airports, borders, in war zones and conservation areas, where their bodies and capacities are tools of labour (see D'Souza, Hovorka and Niel 2020). Porcher, in her work with Elisabeth Lécrivain, is on firmer ground here, asserting that dogs work with humans in guarding farmed animals, and that this provides them with purpose, satisfaction and relative freedom (2019, 113). Unlike farmed animals, dogs engaged in this work may be understood to be *producing* value rather than *being* value. Yet guarding and searching require considerable emotional labour and body work (Hochschild 2012, 7); and however well animals might be kept and their work considered important, this remains a great challenge for dogs that can seriously impact their welfare (Bradshaw 2012, 221). For some animals, emotional labour and body work is part of a commercial setting, for example, in the case of horses used in commercial tourist trailing (Dashper 2019), where the possibility of exploitation of animals through work maybe further enhanced. In response to such concerns, Coulter argues for 'humane jobs and work lives' that are not damaging to animals, people or the environment; and considers dogs to be one of the few species able to undertake such work (2020, 33–34). Guide dogs working with blind people could be seen as undertaking humane jobs. Their work necessitates a deep engagement with dog enjoyment in working with humans while also clearly being work in that dogs have to deploy their intelligence to bridge the gap between the training they received and the situations they are presented with (Mouret 2017).This does beg the question, however, if so few species of nonhuman animal might be able to engage in humane jobs working with humans, is work a conceptual category with traction in securing improved outcomes in the lives of animals? Animal companions are usually seen as matter out of place in workspaces (Charles and Wolkowitz 2019), because as companions, they are not assumed to be engaged in work. Perhaps, however, there are some kinds of human work in which animal companions, and particularly dogs, might be engaged—the mostly privatised work of care.

HUMANS, DOGS AND CAREWORK

Care can be understood as 'everything that we do to maintain, continue and repair "our world" so that we can live in it as well as possible'; it is about taking responsibility for, and being concerned for, the well-being of others (Tronto 1993, 103). We can understand this as the reproductive labour of care which makes lives 'livable' (Puig de la Bellacasa 2012), and ecofeminist discussions of care are instructive here.

Feminist approaches in animal studies have argued that there is a gendered ethics of care which links women to concern for other animals through everyday practices and relationships (Donovan and Adams 2007; Gaard 2012; Deckha 2012). For Mary Mellor, the division of the human Western world into feminised private and masculinised public spheres involves 'placing on women the major responsibility for nurturing and caring' (1992, 251). While the public face of capitalist production takes as its premise 'a sphere of false freedom that ignores biological and ecological parameters', the nurturing private world 'which has its material base in women's time and work' remains invisible (1997, 173–74). The strands of care theory within feminism, ecofeminism and feminist animal studies have tended to focus on care as a human practice that, combined with a critical politics, can enable more positive futures. However, Puig de la Bellacasa asks, 'What does care mean when we go about thinking and living interdependently with beings other than human?' (2017, 13). This suggests the possibility of intraspecies carework in multispecies homes, that animals care for humans.

Coulter (2016c) defines animal carework as comprising tasks and activities involved in taking care of others physically, psychologically and emotionally. It can involve people caring for animals, animals caring for other animals and animals caring for people. However, if we do think of animals engaged in carework, it is often professionalised care that comes to mind, such as Pets-As-Therapy (PAT) dogs who will have undergone behavioural assessment in order to work in schools, hospitals or care homes, or as service dogs providing assistance to people with disabilities (PAT 2020). The demands of such carework are so intense that some suggest this is not an acceptable use of animals (Zamir 2006); certainly, 'the possibilities for exploitation are very high' (Donaldson and Kymlica 2011, 140). In looking at the role of PAT dogs providing calming encounters for stressed students on a UK university campus, Nickie Charles and Carol Wolkowitz (2023) draw on Nicole Shukin's (2013) concept of 'feeling power' to explain how the capacity dogs have for loving attachment to humans is used by PAT and universities to secure less stressed and less complaining student subjects. The animals themselves are volunteered as passive subjects willing to respond to extended periods of petting by people they do not know, in order to make those people feel better. The work of the humans involved is feminised both in terms of personnel, and in being a performance of 'caring' labour. There are distinct power dynamics at play here, for while the dogs' owners seemed to benefit from their voluntary work, the benefits for dogs are dubious, with some disengaging or withdrawing and all dogs finding the emotional and body work exhausting (Charles and Wolkowitz 2023, 11, 15–16).

These dog therapists, working at the behest of their owners and living a life mostly as a companion, are on the cusp between what is often understood to be work and a role which is usually not seen as work—that is, carework in the home. However, in the context of home, dog companions are required to engage in emotional management work on a daily basis, in a similar way to working dogs in guarding, searching and therapy, such that we do not understand the effort involved for a dog in the role of pet until they do not cooperate. We see evidence of this, for example, in the distress and anxiety of so many dogs at being left alone, a condition which affects an estimated eight in ten dogs in the UK and has led animal welfare charities to emphasise the significance of the problem and strategies to tackle it.[3] This leads Haraway to suggest that being a 'pet' is 'a demanding job for a dog, requiring self-control; and canine emotional and cognitive skills' (2003, 38). Most companion dogs companion now live tightly circumscribed lives and 'are expected to be simultaneously better behaved than the average human child and as self-reliant as an adult' (Bradshaw 2012, xx). In Hochschild's (2012, 7) formulation, the private management of human emotion was referred to as 'emotion work', while the management of feeling in order to create a specific facial and bodily display sold for a wage in a public setting was 'emotional labour'. Dogs living as companions are required to manage their emotions in both public and private contexts, and to a high degree. In light of this, dog companions can be understood to perform voluntary and human-tasked carework. As we saw earlier in this book, of all species, dogs have been understood to bond with humans and understand us 'to an extent no other animal can match' (Bradshaw 2012, 170). As a result, dogs are seen to be able to 'continually assess the people with whom they live, physically, intellectually and emotionally, and then proactively and/or responsively provide care of various kinds' (Coulter 2016a, 63); although, she does not evidence these claims empirically. In some of her dog writing, Haraway has exemplified how her dogs (Cayenne and Roland) read human behaviour and situations and positively intervened in caring for her mother-in-law ('K') suffering with dementia:

> Cayenne slipped her body between us . . . and snuggled onto K's lap. . . . Every chance C got, she licked K's face, then pressed her head against K's breasts . . . Roland, meanwhile, had his head inserted between me and K's lap, putting his

> head on her knees along with my hands and pressing firmly against her body as best he could. He also would not budge until K was calm. (2008, 181).

As this excerpt illustrates, the effects of caring are 'brought close' in situations when 'thinking about and with others is understood as *living* with them' (Puig de la Bellacasa, 2017, 17), and below I draw on the ethnographic material to evidence claims made about the carework undertaken by both dogs and humans in multispecies households and to illustrate the different kinds of carework tasks which dogs and their humans practice. This will raise questions that are both interesting and discomforting, such as whether dogs kept as companions—unable to choose where and how to live—are working when living with humans, and if an important element of this work is carework. The stories people tell of their lives with dogs include anxieties about the impact of human work on dog lives, and the carework people undertake for dogs. However, the question the chapter also asks is what might dog work look like, and who is caring for whom, and how?

Empirical research has already demonstrated reciprocal carework relations within situations of pressure and crisis. Dogs can provide motivational support, companionship and physical protection, generally (Irvine and Cilia 2017), to those who live alone or are isolated (McConnell et al. 2011), or those confronting domestic violence (Fitzgerald 2007; Flynn 2000; Laing 2021a; 2021b; Taylor and Fraser 2019). Leslie Irvine's work with homeless people found that dogs and cats were seen to provide motivation to confront mental illness and/or addiction, prevent depression and suicide and to enable homeless people to both endure their lives and to change them (Irvine 2013b). Such animal companions were 'lifesavers' and 'lifechangers', through mutual relations of what was understood to be unconditional love, commitment and an exchange of care (Irvine 2013a, 21–24). The participants in my ethnography were not as vulnerable those in such studies, yet evidence of the lifesaving and changing role of dogs was still to be found.

Carework by and for animal companions is a challenge for feminist understandings of care, however. Within feminism, reproductive or domestic labour, or carework, has been used to describe the set of caring tasks, traditionally performed *by* women for minimal or no wages, that maintain and reproduce life, both daily and generationally *for* humans only (see Weeks 2011). In the 1970s and 1980s, left-feminist writing on reproductive or domestic activities tended to use the terms work and labour interchangeably, and considered the role of such labour to be functional for capitalist relations (Dalla Costa and James 1973; Federici 2004) and gendered social reproduction (Barrett 1980). Such work involves caring for oneself and others (children, elders), through a range of specific tasks (cleaning, shopping, cooking) and species reproduction (bearing children; Hester 2018, 344). These are mundane, everyday tasks which are vital for life (Fraser 2017). The impact of reproductive labour on women has been held to lead to boredom, isolation and depression among those domestically confined (Oakley 1974), and to tensions between the demands of paid employment and reproductive labour (Drobnič 2011). The understanding of reproductive labour in left-feminism has been human-exclusive, but my ethnographic analysis suggests that caring for animal companions in the home might also be understood as caring or reproductive labour, and that the tasks involved have strong similarities to the kinds of everyday practices long identified by feminism.

Ecofeminist approaches—which incorporate what I have called 'feminist animal studies' (Cudworth 2016a; Cudworth, McKie and Turgoose 2023)—have been an important counter to the humancentrism of much feminism. For Mellor (1992, 251), transition to more ecologically sustainable lifeways requires the reappraisal of the importance of feminised caring work that attends to bodies and the demands of 'biological time.' Further, there is a gendered ethics of care which links women to concern for other animals through daily practices. Lori Gruen (2015) argues that in the everyday work of care, a form of perception emerges, 'entangled empathy,' involving both emotional concern and practical caring labour. Everyday interrelations with other animals mean that we are drawn into caring about and for them. The *actual* labour this involves, however, has been little discussed, bar the meticulous documentation of care in personal accounts, for example, for dog companions through aging and death (Pierce 2012). In addition, the discussion of care in ecofeminism has tended to focus on care as a *human* practice. However, I will draw on my ethnography to argue that such ideas of reproductive and caring labour can apply to dog companions engaged in care for humans. This suggests the possibility of intraspecies carework in multispecies homes, that animals also care for humans. Analysis of the data also illuminates the range of specific tasks undertaken by both humans and dogs, in caring for others.

STORIES OF WORK AND CARE FROM MULTISPECIES HOMES

The ethnography indicates three ways in which work and care shape the lives of humans and dogs. First, the ways in which formal paid work undertaken by humans is open (or not) to the inclusion of dog companions and the ways it impacts on the life of the multispecies home; second, the carework undertaken by humans for companion dogs; and, finally, the carework dogs undertake for humans.[4]

Working the Dog

Only five dogs in the sample were regularly taken to workplaces: a book binders, a charity fundraising office, a solicitor's office and a clinic offering alternative treatments such as acupuncture. For some dogs, walking to an office and being expected to sleep under a desk was what was required. While this meant that dogs are not left home alone, a dog is not expected to actively participate and some were unable to adapt to this passive role:

> I took her to the office once but . . . she just wouldn't settle. There's too many people around and too many computer wires to gnaw. (F, London)

Other dogs were permitted more of a free reign within the workspace. A dog in a small office was able to behave as she wished without this being seen as problematic:

> She'll lie with her arms around me and hug me, even at work. I'll be talking to a client sitting on the sofa, and she's behind me just hugging me. (F, London)

In just one case, taking dogs to work developed into a specific ongoing role so that work outside the home becomes part of shared life:

> Well, the first time I took Winnie to work was because I was feeling a bit worried about her being in the home all day long when I was recently separated. And she walked into the clinic and her nose went up into the air and she said, 'Right, what's my job here? Ah, I know, reception therapist'. She was fantastic . . . because when new people come into the reception area they look at each other going, 'Mmm, I'm not quite sure how to talk to you'—and Winnie broke all the ice and everyone started to talk. . . . She'd come into the room when I was treating people and if they were doggy people she'd go up to them, if they weren't doggy people, she'd sit to one side. . . . People tell me, 'I only come here for the dogs, I don't come for you'. (F, Leicestershire)

Most dogs were left at home when people were at work. Only five interviewees were not in any kind of employment, and about a third worked part time. Work was a cause of anxiety for most interviewees, and some had moved from full- to part-time working even when this had proved a financial burden. The first interviewee below, for example, was highly anxious about the affordability of veterinary care, but more significant for her was the need to be able to spend time with dogs:

> I used to work for a top salon in London, and I used to do ten to twelve clients a day. Then I got dogs, and something had to go because I couldn't do the job and have that kind of career with dogs; and the career went because I couldn't

> manage to do both properly . . . I'll work today, but I'll be round the dogs to take them out. I won't be out more than a few hours . . . I couldn't do a full-time job 'specially now that they're getting older . . . a job would not accept you saying, 'Sorry, I can't come in today because I've got an appointment with my vet and he's [the dog] having an operation'. They just don't accept that. So, I work for myself now, yeah, just so I can work my life around my [pause] my animals. (F, London)

> I used to teach full time. I just don't think I'd be happy now, rescuing a dog and then just leaving them for nine hours, I just think it's too long. Or [laughing] maybe I've just convinced myself that I don't need to be at work because I need to look after them. Now I work in the afternoon, part time. (F, London)

> I only work part time, and I'd like to work longer hours now. But I do think about Billy, 'What about Billy?'. (F, London)

In some cases, such as the woman above, people felt they could not contemplate a return to full-time employment even if this were desirable, while others felt that they needed to be working fewer hours. Some were conscious of their dependence on family members and friends providing day care for their dogs. Employers were reported as unsympathetic in terms of making allowances for the illness or death of dogs and were generally hostile to dogs being brought into workplaces. In one case, an interviewee who was a social worker encountered a fraught situation in dealing with a client with a history of domestic violence living in sheltered accommodation (where the keeping of animals was banned) who had acquired a puppy:

> I said to [the client], 'Would you like me to help you with this and find him [the puppy] a home and take him away?'. And he said, 'Yeah'. So, I phoned my manager and she got very confrontational, and I've had a lot of problems at work over it. Well, eventually I took him [the puppy] and she said, 'Bring it back tomorrow', and I didn't. I said, 'I'm not bringing him back because you've not put anything in place to protect this puppy or the tenant that has the puppy. What if he hurts that puppy? You know, it's highly likely with his history and all the people that he's hurt that he might hurt this living thing and you've done nothing to think it through'. (F, London)

For this interviewee, working practice directly challenged her carework for animals and revealed the challenges of dealing with multispecies issues in a working environment. The impact of human work was considered to be the 'worst thing' about living with a dog, particularly when people were working full time and felt guilt that dogs were left for too long. However, these interviewees were often juggling schedules in order that dogs are not left excessively:

> She's never left longer than five hours, and we both can work from home so during the week she'll be with one of us for stretches, and [name of partner] can take her into her office. (F, London)

> between us the dogs aren't left very much. At the moment, he's [her partner] doing two or three days in the office and the others at home . . . if we've got something and we're both out, of course we take them out and wear them out and they might be left seven hours, but that's really a one-off. (F, London)

Not only was concern about the balance between paid work and carework for dogs an issue for my interviewees but this tension was reported in the lives of others too:

> So many of my friends say, 'Oh, I'd like a dog', but they know that they couldn't because of work. That's the thing, isn't it? When I was working full time, I just couldn't have a dog. (F, London)

Whatever the working hours, the majority of dogs wait at home, to greet, to provide companionship or a walk at the end of a human working day. They have negligible agency in changing this relation, which makes it difficult not to conceptualise pet ownership as an exercise of dominatory power which profoundly disadvantages dogs. However, my interviewees were aware of the unfree condition of dogs and conscious of, and sometimes expressed guilt about, the length of time dogs may be left home alone and could become 'bored' or 'lonely'. There is also acknowledgement of their own limited agency when faced with unsympathetic working environments. The field notes for this project are peppered with encounters such as this below, suggesting that is not only the dogs who might experience lack of freedom but also their humans as they juggle responsibilities and the demands of the working day:

> Her partner is in a hurry, calling from the car that they are already twenty minutes late. She doesn't want to go to work today; it's so glorious. (Field notes)

In juggling working commitments, schedules of partners and care for children, other human relatives and other animal companions my human interviewees went to considerable lengths to accommodate living with dogs in their daily routines. Only those with a high degree of autonomy in their working environment (who worked at home or for themselves) were able to manage the demands of work without experiencing conflict and anxiety. For those who had moved to part-time working to accommodate a dog, the decision was often necessitated by a lack of flexibility and understanding on the part of their employer. The question 'What would make your life better, living with dogs?' generated the same answer for working interviewees, in terms of not working, or working less:

> Not having to work! Yes, absolutely, not working [laughing]. Then you wouldn't have to worry about your dogs not getting enough exercise . . . more time with them—that would just make life easier. (F, London)

There is friction, then, between the demands of paid employment and the demands living with dogs places on those who care for them; and the demands of dog care, I will now suggest, can be understood as reproductive labour.

Carework for Dogs

As we have seen, while ecofeminist care theory has been just that and *theorised* the gendering of care beyond the human, there is little work that has examined the everyday labour of care (with exceptions concentrating on care for elderly or ill dogs). This section evidences the everyday, mundane and embodied practices of carework undertaken for dogs which have close affinity with the reproductive or caring labour undertaken for children, other family members or other humans who people are in caring relations with. There were gender differences in terms of concern for the work involved in caring for dogs and homes, yet interestingly, as found in chapter 3, women were more likely to allow dogs free reign of domestic space and to share intimate space with them than were men. Only one of the men interviewed mentioned the impact of hair or mud on the home, or its relation to reproductive labour. The majority of interviewees in this study were women and most raised the cleanliness of the home as an issue, agreeing that 'you have to clean a bit more because obviously they make a lot of mess' (F, London). In addition to cleaning homes, there are the demands of cleaning dogs themselves:

> You've got to have the time, haven't you? Especially the winter we've just had, every time you go out they get filthy, so it's not just the walk, you've got to come back and get them clean and dry, or as best you can, before you can get on with anything else. (F, Leicestershire)

Perhaps the most time-consuming element of carework for dogs is walking and arranging this around work and other commitments. This was a key feature of how my interviewees organised their days, as we saw in chapter 2, and as these accounts suggest:

> It completely changed my life. . . . I'd make sure that I got up early to walk her in the morning and, um, I'd be thinking about her at the end of the day, having to get back. (F, London)

> I leave [for work] at half six, [name of partner] gets up and walks her, just a short walk up and down the field, whatever, just to give her a stretch, and then I'll walk her in the afternoon, or if I'm working later in the day, I'll go out and give her a long walk in the morning, and then [name of partner] will give her a shorter walk in the afternoon, so you know [we] just work it like that, really. (F, Leicestershire)

Such a level of commitment and routine, however, was nearly always seen as something positive. The time involved in walking dogs 'properly'—which interviewees collectively understood to be for quite a long time (forty-five minutes plus), not on roads and usually with time off the lead—was seen as something to be celebrated, as we saw in chapter 2.

> I get up at six o'clock and walk the dog and that is just lovely, lovely to be out, lovely to see the sunrise, nice to be active before work, clears your head, erm, you meet other dog walkers. (F, London)

> Well, you have to go for a walk every day . . . but I'm quite happy to have that responsibility. (F, London)

For some, carework for dogs was understood to be work and described as such:

> Dogs are hard work. They need you, they need your attention, they need your love, they need to be with you . . . they need to be looked after; and sometimes you just have to throw everything up in the air and deal with everything you can to help them . . . you know, that that is your priority. (F, London)

For those with younger children, carework for dogs and for children became part of the same issue around which formal work had to be organised, or 'worked':

> it has a massive impact how I work my work so that I can pick up [name of son] primarily, to save on child-minding fees but also so that Sam is not left all on his own all day because you can't do that. So that's a massive [pause] that is a big thing. I would have a different job or y'unno I would work differently if we didn't have Sam, definitely. (F, London)

For another, who clearly understood carework for dogs and children as being similar in terms of time and commitment, formal working was not possible:

> I get up at six in the morning to take Alfie for a walk before I take the kids to school . . . I ain't got time to work. I've got to look after the kids, I'm constantly

> back and forwards to school and back and looking after Alfie and the guinea pigs—that's a full-time job. (F, London)

Carework for dogs has similarities with care for children—cleaning bodies and feeding and sometimes preparing meals. For a small number of interviewees, the work of care for children was a need, and something that dogs might fulfil as children age:

> When the children are older, you still go on looking for something to look after, I think. It might not be children so it would be a dog. (M, London)

For those who have had children, similarities were also drawn between the social elements of dog walking and walking children to school and spending time at the school gates. In both cases, these were seen as a necessary activity, part of the domestic routine but also one that brought pleasure, for the most part, in being social activity outside home. Also overlapping with the kind of carework involved with looking after children are requirements not only to play with a dog, but to spend time with them and to take account of what might matter in the life of a dog:

> I feel very privileged to have a relationship with a dog. I think it's that . . . people might not necessarily understand in as much as if you're meeting people [after work] they say, 'Oh can't you just walk your dog?' But it's not just about going home, taking her for a walk and then leaving her; because there are [pause] oh [pause] I can't explain it, it sounds really stupid, you know, once I'm there I like to spend time with her 'cause I think that's what she likes, really. (F, London).

> I know that she really loves forest, and I really like to go to the beach. And as I'm driving to the beach, the car seems to take me to the forest. And I think, 'Actually, who am I going for a walk for?' Although I really want to go for a walk, ultimately, you know, it's for her. (F, London)

> Even when we're here [home], we're working or we're busy, whatever, and you know, I do think she probably gets bored. So, you have to think of that, add in some activities that she likes, things that we do that she can join in with. (F, Leicestershire)

While cleaning up after dogs was spoken of overwhelmingly by the women in this study, the labour of walking was discussed by both men and women, and a number of men made explicit comparisons between caring for children and grandchildren and the affective labour involved in caring for dogs. This was seen as distinctive compared to carework for other animal companions living in the home. For example, one commented that having a dog was like:

having children, I think, a similar kind of responsibility, not quite but, you know, it is similar. Very different to having a cat. (M, London)

Some of those working full time used doggy day care; either a set up where large numbers of dogs go to a day-care centre (in the rural study site, dogs would be collected and returned home by one popular local centre in a purpose-designed mini-bus) or where dogs are dropped off or picked up by dog walkers either just for a walk during the day or for all-day care. This was usually seen as a very positive experience for dogs:

She loves it there so much that she always wants to get in the van if she sees it going [a]round the village. We even send her there one day a week even if we are at home all day, so she doesn't miss out. (F, Leicestershire).

Labours of exercising, feeding, playing with and entertaining dogs, of spending quality time with them and providing love and affection are comparable to the labour involved in care for children, as is the organising of day care for dogs and dog-sitting services. Similar to parenting and elder care, were periods of intense carework with anxious dogs, rescue dogs who had experienced trauma, puppies, older and ill dogs; resulting in human carers having disrupted sleep and routines, being unable to work, go on holiday or leave an animal alone.

Yet despite the level of such demands and their emotional toll, such carework was undertaken without regret and a number of interviewees suggested that living with dogs was often more rewarding than living with children in terms of the affection dogs provide and the relative independence of dogs. In the excerpt from the interview below, the woman's speech speeds up as she catalogues the extra work involved in living with a dog until she shifts both tempo and focus, perhaps provoked by the sight of the dog on the sofa, abandoned to pleasure, to conclude that a life with dog-related labour is also a life less empty.

I think life's a lot easier without them because you know you haven't got to be hoovering up [vacuuming] or sweeping or mopping or cleaning; and also, you've got time on your hands because you're not out a couple of hours a day dog walking, so life is easier. But I think it's emptier without one, so you know we have always [pause] oh look at Ruby [sleeping on a sofa, on her back, with her legs in air]. I have to say, happiness is dog shaped. (F, Leicestershire)

These specific forms of labour undertaken by human carers for dogs should be understood as caring or reproductive labour. This comparison suggests the need for further feminist research into the impact of carework for nonhumans, and the ways this is bound up with, and also different from, normative feminist understandings of reproductive labour and carework. This is a matter to which I return in the discussion below, but first, let us consider the final element of labouring with dogkin—the possibility that dogs undertake carework in domestic relations with humans.

Carework by Dogs

While all the tasks of care that dogs undertake involve emotional labour, perhaps the most significant is the management of boredom and anxiety when left at home, particularly for long hours when people were at work. The impact of human work on dog companions is rarely discussed in literature apart from the ethical issues raised by the confinement and leaving of dogs (Pierce 2016). Only five interviewees were not in any kind of employment, and the majority of dogs waited at home, to greet, to provide companionship or a walk at the end of a human working day. Having a dog in the home to whom people returned and who is 'always so pleased to see you' and 'greets you' was mentioned by nearly every interviewee. Thus, an important role that dogs are expected to fulfil for their humans is simply to be there:

> she's always here, even if you're in the house on your own, you've got somebody. I talk to her all the time, and I think they're a real comfort and, if you're feeling a bit down, they can usually cheer you up one way or another, can't they? I just love having her, really. It's funny, I just [pause] yes, I wouldn't be without her for the world now. (F, Leicestershire)

> an unconditional love, you know, they never ask questions, or they are, umm, they're always there and even if you have a bad day, the dog is not going to question you, or make it worse [laughs]. (F, London)

This role of simply being there, waiting for humans to return to the home is a cause of anxiety for very many dogs. In the UK, there are estimates of up to 85 per cent of dogs suffering separation anxiety when left alone, 50 per cent of whom are managing their emotions so effectively that their human companions are unaware there is a problem (RSPCA 2019). The isolation and boredom of life for a dog confined—however apparently comfortable the home—is a matter which surely resonates with some very established feminist concerns (Oakley 1974). As suggested earlier, if we consider emotional management to be labour, then we can consider the expectation to wait and to greet to be a task of care. Dogs have negligible agency in changing the conditions of their confinement, however, my interviewees were aware of the unfree condition of dogs; conscious of, and sometimes expressed guilt about, the length of time dogs may be left and could become 'bored' or 'lonely.'

In addition to waiting and greeting, an important related task dogs are expected to fulfil is to provide company as a support for their owners. The importance of human-dog communication is emphasised, whether it is the nonverbal communication between dogs and humans as they greet each other when a human returns to the home, or the presence of a creature which responds to what Clinton Sanders calls 'self-speech'. Most people, both male and female, talked to the dogs they lived with inside the home, and they understood this as a genuine exchange—dogs are both company and companions because they share communication in home space:

> People who have been in my house they go, 'Who are you talking to?', 'I'm talking to my dog'. It's like having company all the time, isn't it? (F, London)

> They're company, aren't they? I get more sense out of the dogs than I do my husband [laughs]. He'll say, 'What did you say?'. I say, 'I'm talking to the dog' [laughs]. They know what I'm doing, don't you [directed to the dogs]? You know where I'm going, yes. (F, Leicestershire)

> My oldest son is twenty-six and the youngest one is sixteen; they're not ever-so-keen on talking to me, really, but the dog is so happy when I talk to her so that's really nice. I sing to the dog and I'm often not aware that I do. (F, London)

> If the phone went and I'd be talking, she'd think I was talking to her, so she'd start conversing back. The louder I got, the louder she got, so then I'd say, 'Excuse me, I just have to go and put the dog outside because I can't hear what you're saying' [laughs] . . . she was a wonderful character. Yes, she was a real soulmate. (F, Leicestershire)

Interviewees commonly expressed the view that their dogs understood (at least some of) what was said and that hanging around together was companionship. Most of the women interviewed, in particular, commented on the tactile presence of a dog:

> a living thing that I can hug and stroke. . . . I just love that feeling that I'm looking after her and caring for her. (F, London)

> I think it's a really lovely thing to have a dog in the house. You can stroke them, and you know they follow you around and look at you and they understand everything, they do, they absolutely understand. (F, London)

The presence of dogs was often described as 'relaxing', and perhaps the ability of dogs to live in the present and enjoy the moment is what enables them to intervene in human interactions and exert a calming influence.

> I always feel so relaxed and happy with them. (F, Leicestershire)

> They're very good therapy—when you've had a stressful day and they come and make a fuss of you. They put their head up and you give them a little stroke and it's just relaxing, isn't it? (F, London)

On occasions, this calming influence takes on a more significant role as dogs actively intervene to mediate in relations between human occupants in the home:

> My marriage was falling apart, but what Millie did was she got . . . rid of a lot of the tension in the house . . . I remember just situations and saying to myself, 'It's really good that we've got a dog', you know, situations where there was arguments or possible tension, and the dog would take away from that somehow by just jumping up and playing or distracting the vibe in the house. (F, London)

> Sam hates it if we are arguing in the house. Obviously, that doesn't happen very often [laughing] so, umm, if we are shouting, he really hates it. He will go from one [person] to the other and do a little bark, yep, he hates it and looks quite like [impersonating the dog], 'Oh my God, please sort it out'. I don't know, it just adds another dimension, doesn't it? (F, London)

> It's that third person in the house, it's that third ear. I've just thought about that, that is true, it's someone to communicate with. I suppose it doesn't really have to be a dog, it can be any animal, it could be the cat, but dogs are much more *with* us, and I think that it helps with any tension at all in the house, you can communicate with the dog [dog barks]. (F, Leicestershire)

The presence of a dog can help smooth tensions because dogs can be a conduit for speech between humans, have a role of confidant or act as mediating presence in the home, as in these examples above. In some cases, dogs are 'lifesavers and life changers' in situations of relationship break up or serious illness:

> she was the one that made me go to the doctors and told me that I had my breast cancer because every time I sat down, she used to get her nose and go like that to me [nudge her armpit] and cry, I couldn't sit down without her being there, nudge, nudge and crying. . . . And it was [pause], so it was quite sad when she went [died]; she was my little angel. (F, Leicestershire)

> I used to have therapy . . . and it was like, 'Why? Why is that so important? Why is that relationship with your dog so important?'. And it was because it was very, I suppose, very unconditional. (F, London)

> She says, 'Freya always looked after me' and tells of her relationship break up, and how the dog would fetch the box of tissues when she sat on the sofa in tears—'She was never trained to do that, she just did it'. (Field notes)

> My brother is less stressed now he has his dog; his wife noticed the difference straight away. . . . He's come off all his pain killers. He tried that before and it didn't work, but since he had the dog, it has. Because the dog is quite attached to him and lies over his feet. (F, London)

> So, he's helped me, I guess. You see, I feel a little obsessional today, I don't know why, something at work, and I find I come over here for an hour or two with Ronnie, and it helps. (F, London)

Most interviewees were keen to emphasise that dogs understood human emotions and that they reacted to them; in particular that dogs would seek to comfort those who were upset. This notion of dogs being able to read humans and as emotionally reactive is expressed in various ways, effectively surmised by this participant:

> I feel they're quite intelligent. I feel that they respond to how you feel. They feel your emotions . . . if you are a bit sort of upset or something, I think they feel that . . . and they want to be tactile and touch you, they want to be close to you. You know, if you were upset or you were crying, they'd probably come up to you and nuzzle into you. (F, London)

Interviewees overwhelmingly understood dogs to be engaged with them in relations of affective reciprocity. As their human companions suggest, dogs engage in various forms of carework: they greet and welcome people home, provide company and affection, they may endure but also seek tactile encounters, and this can constitute active intervention in response to human emotions. By choosing to spend time with humans, in putting up with being talked to and sung at, stroked and cuddled, dogs provide company and comfort to their human owners. This is likely to come at a cost—the boredom and anxiety of being left alone, a lack of personal space and emotional demands when dealing with intrahuman tension in the home.

MULTISPECIES CAREWORK AND THE POSTHUMANIST IMAGINATION

Coulter (2016a, 146) argues that recognition for animal work is critical in changing how animals are understood as positively contributing to society and to consider how animal work can promote interspecies solidarity.

However, there are questions as to the extent to which recognition of labour might really have transformative potential for animals. As we have seen, there is research attempting to suggest that recognition of animal labour in agriculture might transform modes of production in the interests of animal and human welfare. However, the notion, for example, that dairy cows are working stretches the category of work considerably and suggests this might be a case of over comparison, where we 'compare animals to humans so that we care about them' (Deckha 2018, 227). A small number of a very limited range of species (horses and other equids, camels, seals and sea lions, rats and dogs) might engage in work as more conventionally understood; and for many of these creatures, this is conscripted labour rather than 'humane jobs'. This suggests that work is a category with limited traction in securing emancipatory outcomes.

There needs to be a more critical edge to the discussion of animal work that includes a critique of the work-centrism of many contemporary societies and which 'aims at reducing unnecessary labour: enabling flourishing by diminishing work' (Wadiwel 2020, 200). For Kathi Weeks (2011), endorsing productivism and the work ethic as a social good, leads to a range of injustices, exclusions and exploitations. Sue Donaldson and Will Kymlica (2020) note the intense humancentrism of discussions of 'post-work society'. They argue that work, as currently conceived, is unrealistic (given future technological predictions), underpins social hierarchy and is ecologically unsustainable (2020, 213). Donaldson and Kymlica suggest a post-work ethic which stresses the importance of the common reproduction of life and participation in the commons would be a more fruitful route for the inclusion of animals. Rather than shoehorn animals into work, we should think of our relations with them as a 'good fit' for the development of post-work society for many

> humans who spend time with animals find this time intrinsically well spent, as well as a source of renewal, pleasure and joy. These are not superficial extras, or side benefits, of living with domesticated animals. They are the fundamental glue of society, upon which the complex relations of cooperation, negotiation and reciprocity depend. (2020, 225)

Certainly, a reconfiguration of human work enabling more time with animals would significantly improve the lives of my interviewees, and most likely their dog companions.

Coulter (2016a) is right to suggest that carework is probably the most promising arena for thinking about animal labour. This is also an area where we may see an element of reciprocity when we think of dogs and humans as our companions in the home. Feminist arguments for a move towards caregiving *societies* are important here (Tronto 2013) and might be open to the inclusion of animals such as dogs undertaking caring labour for members of their own household and for others. The labour of caregiving overlaps, as we saw in chapter 5, with kin and friendship, which are productive categories when considering human relations with dog companions (Haraway 2016a; Charles 2014). Such work suggests that there is interdependency in these relationships running alongside the significant power differentials in a human-centred social world. The work of care, by human and dog companions, involves the pleasures and labours of spending time. Care is both relational and situated (Cuomo 1997, 130). Caring labour is emergent through everyday '*living with* others', and as the data analysis from this research suggests is to some degree 'agentially distributed' and as existent in 'a thick mesh of relational obligation' (Puig de la Bellacasa 2017, 18–22).

One interviewee, experienced with rescued dogs who had been in abusive situations, remarked:

> One of the things that really gets me about dogs is how loving and trusting they are, even if they're ill-treated by people. They're quite an amazing creature in that way, brilliant, they'll still love you, even sometimes to their own detriment. (F, London)

The labours dogs undertake in the home are often labours of love. In a society where human needs and interests predominate and the demands of work practices in capitalism fundamentally shape human lives, dog companions may indeed love us 'to their own detriment'. While the public world of human paid work affects both humans and dogs, and often to the detriment of both, the labour of care in the home offers a space enabling more convivial relations. I understand the carework of humans and dogs as a set of emergent labour practices which come to matter in *living with* each other.

How might we conceptualise carework for and by animal companions such as dogs? First, it is useful to look outside feminism for a less human-exclusive concept of labour. Rather than use work and labour interchangeably, Hannah Arendt makes a clear distinction. She understands labour to be those activities necessary for 'life', and as distinct from work (1998, 85–105). Arendt's labour is not human-exclusive and alludes to reproductive/caring practice. The terminology of reproductive/caring *labour*, as per the Arendtian conception, I consider to be a better fit for capturing multispecies care. For decades, Western feminism struggled to make visible, and account for, a whole range of everyday practices, behaviours and interactions that have been understood as unremarkable, assumed to be about love or emotional connection and obligation. These practices, naturalised and strongly gendered, have been referred to as domestic or reproductive labour. We might revisit feminist analyses of caring, emotional and reproductive labour in thinking about dog–human domestic relations in (relatively) wealthy societies, such as the UK. While housework produced the subjectivity of the housewife, and the labours of care for children produced the subjectification of motherhood and the cultural motif of the good mother, human reproductive carework for dogs can be seen as productive of the dog owner or guardian, and the motif of the responsible dog owner. Those I interviewed in the course of this project saw themselves and/or certainly wanted to be seen as a 'responsible' or 'good' dog owners.

One of the important contributions of feminist analysis of domestic/reproductive labour, as we saw earlier, was to specify its composite tasks as labour practices. The range of reproductive labour tasks familiar to feminism include cleaning homes and persons, shopping, tidying, training, socialisation, preparing meals and feeding, entertaining and playing. I understand the reproductive labour of dog owners to be very similar and to comprise the following cluster of tasks: providing daily exercise, cleaning and grooming, enabling socialisation with people and other dogs, training, feeding (which will involve shopping and perhaps cooking), alongside the emotional labour of providing attention, love and affection. The reproductive economy of carework has shifted for dogs, and increased levels of responsibility for, and control over, dogs, means that commercial interests have burgeoned in the provision of carework and additional services which require management (including non-medical veterinary services, grooming, day care and walking services). If we consider the reproductive carework undertaken by dogs, there are overlaps, as the cluster of tasks may involve: providing daily exercise, enabling socialisation (with dogs and humans), and providing company, emotional support and affection, play and entertainment.

A focus on this embodied and emotional labour shifts us away from the difficulties in endorsing productivism. It also encourages a posthumanist resistance to the work ethic and the exclusions and hierarchies of productive labour under capitalism. The demands of the latter produce a range of difficulties for dogs and humans as companions. When we hear of family friendly workplaces and ways of working, we assume this means accommodating of the domestic responsibilities of children, or perhaps aging, ill or disabled human family members. This might be extended to include nonhuman companions also, an issue to which we will return at the close of this book. Inspiration for this chapter came from Haraway's provocation that being a pet is a difficult job for a dog, and insights from various kinds of feminisms have shaped my response: a need to recognise the complexity of caring labour as multispecies but also in ways that are reciprocal; to recognise care *as* labour; and to be attentive to power and the darker side of reproductive/caring labour. Rather than shoehorn animals generally into work broadly defined, it is more helpful to consider specific species, relations and contexts and particular forms of work/labour. When it comes to dog companions, intraspecies reproductive caring labour can help us better understand humans and dogs as domestic partners. In addition, a gender lens is needed to revisit issues of affectional slavery and consider how labours of love might be other than this. Feminist recalibration of society in ways that values kin, communities and care, and that enables flourishing by diminishing the significance of work, needs to embrace multispecies worlds and practices.

Conclusion

As readers will be aware by now, my understanding of dog companions is that they are constituted through a relational matrix of anthroparchal domination alongside affectional ties that can be productive and supportive of animal agency. I hope to have reinforced the importance of ethnography in capturing nuance in human relations with nonhuman animals by inscribing daily lives and practices both in and outside of the space of home. The research on which this book is based does raise questions about the inevitable humancentrism of research practices in the social sciences, the challenges of undertaking genuinely multispecies ethnographic work. While dogs and their lives with others have been central to this book, dogs themselves are less present in its pages than I had hoped. This is perhaps inevitable, a result of my reliance on human accounts of lives shared with them. Moving away from an ethnography of sight might be one way of bringing animals to the fore in our projects. Certainly, if we want to better understand creatures such as dogs, thinking about sound and smell and ways in which dogs know their worlds may be key. This would lead to a very different project than the one on which this book is based.

Living with dogs entails relations of both domination and affection, and the ethnographic material this book presents suggests that heterotopia is found in the everyday practices of multispecies living. Dogs have become experts in reading humans—certainly the stories from human participants demonstrate ways in which dogs empathise with their humans, seek to register their disapproval, seek to persuade us or anticipate what we might do. The spaces of muddied living in the lives of people and dogs are ultimately understood to be heterotopic spaces in a nexus of human domination. I will conclude, therefore, by considering what possibilities this offers. How might the heterotopic spaces of muddied living be nurtured, expanded, multiplied? What, in turn, does this imply for the organisation of our humancentric social life? How can learning to live well with dogs in everyday life raise awareness of an array of ways in which we might try to 'live less wrongly' with those we have tamed, commoditised and call friends? The muddying of boundaries between humans and dogs raises questions about what it means to be a human and what it means to be dog. Here, I think through the implications of what can be glimpsed in these spaces of other possibility and what this might mean in terms of practical political interventions. First, however, let us consider the muddied living of people and dogs through the lens(es) of heterotopic space.

MUDDIED LIVING AS HETEROTOPIA

In his well-known essay 'Of Other Spaces', Michel Foucault introduces the concept of heterotopia as a different way of thinking about space. By contrast to the 'fundamentally unreal' or fantastical space of perfection—utopia—heterotopic spaces are real spaces, relational spaces and 'a sort of mixed joint experience' which reveal paradoxes (Foucault 1986, 24). Heterotopias are spaces which open and close, which have distinct qualities but are also penetrable, where incompatible or contradictory spaces come together. Muddied living, I suggest, can be usefully conceived as an area of interlinked contradictory experiential spaces.

Much has been written and debated on the notion of heterotopia. Some have seen the idea as something for which Foucault had limited affection, or which was underdeveloped and incomplete (Soja 1996; Saldhana 2008), but it has nevertheless proved popular. Many who have operationalised heterotopia have used it as a concept of resistance, emphasising Foucault's description of heterotopias as 'counter sites' (1986, 24) or 'anomalous zones' (Arcari 2020, 3) which function differently to dominant norms and practices. Some have seen this as strategically useful for movements seeking to transform human-animal relations in radical and liberatory ways (Socha 2012). This suggests an alignment with the space of perfectability implied by utopia, and there are elements of overlap between utopia and the practices of prefigurative politics or radical speculation. For example, Ophelie Véron (2016, 769) argues that veganism is a prefigurative co-constitution of activism and/in everyday life, so that the everyday space becomes a series of what she calls 'hemeratopias' (after Hemera, primordial ancient Greek goddess of the day). Alternatively, Michelle Westerlaken (2020; 2021) advocates 'speculative multispecies worlding' by suggesting that an understanding of species multiplicities need to be carried into design processes in order to generate interspecies harmonies rather than reinforcing oppressive relations. Speculative multispecies design work, Westerlaken suggests, can propose other kinds of relations between species, as well as embody more productive ways in which humans and other species already make worlds together. Neither of these prefigurations from critical animal scholars, however, would capture the ways in which the everyday spaces of human-dog lives are contradictory. While there are elements of life in posthumanist households which enable a glimpse into 'alternative pathways that might glimmer beyond the system' (Arcari 2020, 84), currently, dog–human worlds are lived in ways where the alternative pathways glimpsed are decidedly muddy. Foucault himself uses various case studies of heterotopia—museums, hospitals and prisons—and considers these public institutional sites as part of interlinked spatial categories. His examples do not speak to dog–human relations in everyday contexts particularly effectively. So, in making a case for dog–human space as contradictory, I want to run elements of prefiguration and multispecies worlding alongside Foucault's heterotopia.

Véron's hemeratopia seeks to situate prefigurative practices in the everyday worlds of home, work and use of public space, rather than within the potentially repressive intuitions of prison and hospital that Foucault focuses upon. I want to take from Véron the significance of everyday practice. The home is clearly spatial and imbued with meaning. As Mary Douglas puts it, 'the home is the realization of ideas' (1991, 290). Westerlaken's multispecies worlding emphasises how interspecies relations can be 'built in' to the material life of the home, and how being attentive to the experiences of other species can change our practices this is also useful. However, both Véron's and Westerlaken's conceptions are too utopian for thinking through what happens in everyday lives with dogs *as they are currently lived*. Westerlaken (2020, 109) is careful to make clear that multispecies worlding assumes no final position, but is an evolving process that remains prefigurative. However, as Debra Benita Shaw points out, Foucault's examples suggest that he sees heterotopia as functioning 'to sustain the hegemony of dominant institutions' (2018, 103). In this sense, they are not a glimpse in the mirror of possible utopia, nor are they a prefiguration related to radical politics. Rather, they are a reflection which embodies both contradiction and possibility. This is how I am using heterotopia here. While there is much that those living with dogs see as positive, in a human-dominant, anthroparchal social world, dominant practices of species hierarchy are the frame within which dogs (and humans) are constrained. In Véron's hemeratopia, vegans occupy spaces which are attached to, and shaped by, anthroparchal hegemony, but within spaces of home and of activism, there is utopian prefiguration. In the posthumanist home shared by humans and dogs, species difference is a difference of power; and human power over dogs is played out (and sometimes contested) daily. Contradictions abound, rather than utopian prefiguration.

For those who have used heterotopia, subjectivity has been a crucial element as heterotopia has different meanings for those (people, it is presumed) who engage with them. Kevin Hetherington (1997) suggests that heterotopic space will 'always have multiple and shifting meanings for agents depending on where they are located within its power effects' (1997, 51). As Shaw further notes, however, they are sites of 'concentrated subjectivisation' while also being sites where this can be contested (2018, 104). This point is crucial. Whose ideas are realised in the space of home shared by humans and dogs? Heterotopia implies difference and challenges in the kind of space in which we may feel at home, but ultimately, as humans, we hold the power in our relationships with animal companions. While people enact the subjectivity of the dog owner in their relations and practice, dogs are subjectified as human companions. By preferring a (perhaps gloomy) reading of Foucault's heterotopia, I do not mean to imply criticism of prefigurative politics; elsewhere I have endorsed it (Cudworth and Hobden 2018, 150–56); however, while posthuman homes and communities *might* be part of a prefigurative, transformative politics of multispecies worlding, the dog owners in my study are far from contesting the status quo in the here and now.

I choose heterotopia to describe what is going on in dog–human worlds because it both signifies the institutions and practices of species domination that frame lives of dogs and their people and are reproduced in everyday practice, while also enabling discussion of spaces and moments of contestation. Some of the areas of everyday practice covered in this book are more open to contestation than others, and some are more dystopian than places of heterotopic encounter. How then, can heterotopia be glimpsed across the practices covered in this book?

The practices of living with dogs involve the knowledge of and familiarity with outside spaces. Parks, streets and homes, and often the homes of other people and dogs, are the linked places where daily routines occur. They are also spaces of difference—walking with a dog is a very different experience to walking without a dog. For dogs themselves, walking without a human may be an altogether different experience but is one which, in a country such as the UK, we rarely ever see. Species difference may also mean that the familiar is often represented as strange. In the 'edgeland' space of Walthamstow Marshes, my field notes recall:

> An afternoon walk in the sunshine. Kevin enjoying himself and ahead of us. I'm a bit behind with things and take a book with me to read while walking. I can usually manage this without tripping up—my feet know the route very well. Some bloke tells me off when he passes me on the boardwalk—for reading when I should be enjoying the weather, my dogs, my walk. I smile and say something banal rather than telling him to fuck off—it's too nice for a row. I look up as I come off the boardwalk and down on to the Lammas field and stop in my tracks. There is a bunch of dogs, on their own [that is, without humans], lying together on the grass. They are different breeds, sizes and look to be different ages; there are about eight of them. A few of them I recognise as coming

> from the boats [moored on the River Lee]. I want to watch them for a bit so divert from the path across the field and go around the edge, walking slowly. Gypsy and Kevin completely ignore the dogs. After five minutes or so, one of them, a large and older looking dog, gets up, has a shake and ambles towards the boardwalk. The others do the same and follow. It is all very relaxed, and it seems very familiar—at least to this group of dogs. They process, slowly, down the boardwalk back to the river, and, I presume, to the boats they live on. As I write this, I know I will be thinking about this for weeks. Doing their own thing, fantastic. How many dogs in our misnamed 'nation of animal-lovers' ever really get to do that?

This group of dogs showed me something entirely different in a highly familiar space. By being alone (or more accurately, without humans) and being in a group, they contested my idea of dog walking and of what being a dog means. Their difference created a space of illusion, they represented the space as something that was about their imagination, exposing and making visible what I, as someone living with dogs, was unable or perhaps unwilling to see. They made visible the normative unfreedom of dogs and represented the possible autonomy of dogs. They also made me take back my irritation at the man who told me off for reading while walking.

As we saw in chapter 2, the relatively free spaces of edgeland in which dogs walk are spaces of walking packs that are dog and human, and always presumed to be together. The edgelands may only enable partial freedom for dogs, indeed, my glimpse reported above may be that of utopia! However, edgeland spaces are heterotopic in enabling the development of a community that is, by its nature, interspecies. I found that posthumanist communities of human-dog walking packs have certain characteristics. First, there is identification with outside place(s), and a shifting and dynamic emergence of community, particularly as the community is generated, reproduced and intensified through the process of walking. Second, this community is not exclusively human. At one level, dog walking communities consist of dogs and their humans in individual and cross-pack interactions. In addition, the agency of dogs is apparent in the creation of interactions between dog walkers and this is key to the deepening of relations or the absence of any kinds of connections. Both these elements, in my view, are sufficient for such communities to be understood as posthuman. Third, these posthumanist communities are characterised by a relatively high degree of tolerance of diversity. Humans from various kinds of social backgrounds are brought into proximity through the process of dog walking. In addition, there is tolerance of some diversity of human-dog relationships and of dog personality and behaviour and a range of practices of care and respect within these communities, including concern for the health and well-being of both human and canine members, and also care for place. At times, more traditional forms of political engagement emerge from the debates and discussion amongst human walkers and focus on the protection of edgeland space and walking access and the generation of dog-friendly public policy. These communities emerge over time in places but are open and closed and shifting in the way Foucault's notion of heterotopia suggests, and attached to anthroparchal normalcy whereby the movement of dogs and people is restricted, by ideas about appropriate dog behaviour, or pressures on human time.

When it comes to the muddied living in homespace, the dogs in this study were, as Varner (2002) suggests, mostly domestic partners, otherwise animal companions and never 'mere pets'. There were different degrees of intimacy present, but every interviewee had something to say of how their lives, the patterns of activity in their households and the experience of 'home' is altered by the presence of dogs and how their relationships were with particular dogs as individuals. There is not purity of species in the posthuman home but what Tracy Jensen and I have previously described as a 'mongrel domestic' (Cudworth and Jensen 2016) in which human(s) and dog(s) come together in various degrees of spatial, physical and emotional intimacy. In muddying dwelling spaces and domestic relationships, dog companions are active in making a different kind of home. Despite anthroparchy in the public world and abuse and neglect in many homes wherein dogs are kept, for both human and canine dwellers in my study, home *can* be a place where love and care happen and dogs can perhaps express selfhood more authentically as a lively presence attempting to shape their place. Heterotopia in the spaces of muddied living can be found in the contradictory dog and human understandings of the use of space and ways these are both juxtaposed and accommodated. In muddying our dwelling spaces and our relationships, dog companions are active in making a different kind of home—a mongrel domestic place. Yet within the narratives of the 'responsible', engaged and concerned dog owners I spoke with, power makes its presence felt in more subtle ways, for example, in the uneasiness many feel about the status and lives of animal companions often left in the home for long hours by humans leading exclusively human lives at work, in the supermarket or at dog-free leisure events. Here, micro-level data—stories of mundane and routine lives lived with dog companions—fold into the macro, wider social structures and practices of human relations with companion dogs. The posthuman home is not a space of equality, and for some domestic animals captive in the home, it is a place of deprivation. However much loved and worried over, a dog left home alone was still home alone and likely to be finding long hours alone boring, stressful and lonely.

An important space of contradiction was the kitchen of the posthuman home, where bodies of other animals may be fed to dogs, or where these might be prepared for human consumption. For those for whom eating animals was normative and unproblematic—'unrestricted carnists'—issues of an appropriate diet for their dogs were not contested or contradictory, but the majority of people interviewed did restrict their carnism in certain ways. Participants' discussions of their own diets, and those of the dogs they live with, revealed tensions and uncertainties regarding what dogs should eat, and which animals can be eaten as food. For the vegetarians, ambivalence was heightened. This said, the responsible dog owners in this study cared very much about the health, welfare and happiness of the dogs with whom they shared their lives.

The interviewees also had concerns for animals other than their own dogs, for dogs in general, for animals kept as companions more widely, for wild animals and in some cases, for non-companionate domesticate animals. Certainly, there has historically been a tension between concern for 'the environment', more broadly and that for other animals—in particular, domesticates. In their dependency on humans, domesticate animals have been seen as expressions of inauthentic being, lacking agency and autonomy. Yet this is a false division between what is 'wild' (that is, uncultivated, untamed, undomesticated) and what is not. Rather, there is a scale wherein some creatures and their environments are, more or less, more and less 'wild'. Simply because a creature is not domesticated, does not mean it is not caught up in relations with humans, either of some level of dependence on them or perhaps of co-dependencies with them. For example, dingoes have commensal relations with indigenous peoples of Australia, living in close association (see Bird Rose 1992; 2011). Saunara Taylor (2017), along with other feminist influenced writers, has critiqued this conception of dependency. We are all intra-dependent and 'able' and less able at different times and contexts in our lives. The discussion of diet is a potent reminder of the contradictorily entangled relations of species in relations with dog companions. In *The Companion Species Manifesto*, Haraway suggests that 'the task' is 'to engage in a joint dance of being that breeds response and respect' and then to make a wider leap 'to remember how to live like that at every scale, with all partners' (2003, 62). In *Social Lives with Other Animals*, I began with some reflection on my vegetarian grandfather's provocation to my six-year-old self: 'Would you eat your dog?' (2011, 4). There is a serious need to consider what response and respect mean in our wider relations with animals whom we do not consider companions, but with whom we share this world. I would hope for greater awareness of the treatment of other animals by those living with and loving animals, and that the experience of sharing life with another animal in close proximity and intimacy might encourage more people to 'stick to grass', as one of the participants in this research put it.

Participants did challenge humancentrism in other ways, however, and, while seeing dogs as family or kin, were resistant to the notion that dogkin were substitutes for human relations. The muddied intimacies between species played out in everyday relating between people and dogs in the relatively wealthy Global North, and in the lives of participants, both human and dog, I consider is better described in terms of making kin, than being family. My research suggests that the idea of making kin—forging intimate ties and attachments—may work better for dogs and some other mammals and some birds than for other creatures. Certainly, some species were too odd to be kin for my human participants. This leads me to conclude that while I have much sympathy for Haraway's notion of kin-making, we need to be specific and tease out different elements of relationality in our response to various species. People expressed passion and intensity in describing the kinds of connections they had with the dogs-of-their-heart. Love was a term used liberally. Marc Bekoff considers that 'love is the answer' to the fraught relation that humans who are Western and relatively privileged mostly have with the nonhuman living world. He goes on to say that 'we must intimately connect with and love other animals, other humans, and all environments if we are to continue to live with grace and harmony on this wonderous and interconnected planet' (2002, xxi). Alternate cosmologies have much to teach here. Certainly, there is more room for fuller engagement with concepts and ideas from indigenous scholarship about what worlds build liveable worlds and how liveable worlds with other creatures, beings and things might be crafted is the challenge we face (as suggested by Rosiek, Snyder and Pratt 2020; for examples of concepts, see Simpson 2017; Viveiros de Castro 1998).

The loving and caring relations between humans and dogs can also be understood as a form of work. Jocelyn Porcher and Elisabeth Lécrivain's suggestion that for dogs, working is 'mainly living together', because 'whenever dogs are bonding with humans they are working' (2017, 119) can apply to some of the labours companion dogs undertake. Yet as I argued in chapter 6, I do not see work as a productive category for dogs kept as companions, because it tends to become overly associated with the kinds of work humans do. I do think that the labours dogs undertake in the home are often labours of love and shaped by the confluence of affection and power. In a society where human needs and interests predominate and the demands of work-practices in capitalism fundamentally shape human lives, dog companions may indeed love us 'to their own detriment', as one of my participants put it. While the public world of human paid work often negatively affects both humans and dogs, the labour of care in the home offers a space enabling more convivial relations. The carework of humans and dogs is a set of emergent labour practices which come to matter in *living with* each other. Carework needs to be recognised as multispecies and in ways that are reciprocal; to recognise care *as* labour; and to be attentive to power and the darker side of reproductive/caring labour.

What do the heterotopic spaces of human and dog lives suggest for sociology and the social sciences more broadly? The structuring of this book has related to key sociological concepts and matters: theory and methods in the opening chapters; then community; home; consumption, food and eating; family, kin and personal relations; work and care. All of these areas of enquiry have been opened up to study beyond the human, but far more is urgently required. My hope is that this project, in some small way, will contribute to the emergent and increasingly assertive voices arguing not just for the inclusion of other animals in the social but for the active transformation of the social sciences. Charting of the inclusion of animals in sociology, Bob Carter and Nickie Charles (2018) draw on early evaluations of feminism for the transformation of sociology in order to identify four phases. Carter and Charles retain the feminist terminology of gender and women, but in reference to animals, I would describe these as follows: first, an exclusively human era when nonhuman animals were not a research focus; second, critique of such humancentric neglect; third, growth in the number of studies which 'add in' animals; and fourth, full theoretical integration of species into the discipline. Carter and Charles convincingly suggest that sociology is between stages two and three, with a critique of the lack of attention to nonhuman animals, on the one hand, and a burgeoning interest in human-animal relations and the growth of the interdisciplinary field of human-animal studies on the other. They do not presume the smooth progression of integration, however. Certainly, some consider such work to be marginalised and perhaps career-damaging. Rhoda Wilkie (2015a), for example, describes sociological animal studies as 'tainted scholarship' and far from the mainstream, and I do think this remains the case today. I have spent much of my career thinking and writing about feminist approaches to nonhuman animals, to nature and to the nonhuman environment or lifeworld. This has certainly not been a journey of smooth progress either, but one of loops and turns. Feminist dog writing, the wide corpus of ecofeminist work (on other animals and beyond), the expanding field of feminist animal studies have been part of this project, this book and my life. Feminist recalibration of society in ways that value kin, communities and care and that enables flourishing by challenging oppressive and exploitative practices has embraced multispecies worlds, but there is an urgent need for further work and more inclusive institutional and scholarly community practice. Being a feminist social scientist looking at relations with nonhuman animals should not be tainting; and I passionately believe it will be a key element in the future of feminism. In addition, academic institutions and procedures and scholarly remits and practices need to be open to critique and actively respond to posthumanist troubling of the subjects, methods and purpose of research (see Oliver 2021).

BE MORE DOG

In *Unleashing Your Dog*, Jessica Pierce and Marc Bekoff, as we saw in the introduction, recommend that key to living better with dogs is to understand what it means for them to be essentially a captive species, and to enable dogs to behave like dogs as far as possible for as much time as possible. As Mariam Motamedi-Fraser (2024) argues, what we often think of as 'normal dog behaviour' is that which is demanded by humancentric society and is defined as both human-orientated behaviour and acceptable in the limited field of behaviour dogs are expected to exhibit. The 'species story' of a dog as by definition a human companion, she asserts,

> justifies and legitimates disrespect at every level, from imagining that a dog welcomes every human touch to the notion that it is the *raison d'être* of a dog to be loved by, or to work for, or to be friends with, or simply to *be*, with humans. (2024, 236, emphasis in original)

Dogs have an enforced mode of becoming in which relationality is all, she suggests. However, much as the power politics of the species story of becoming dog is highly problematic, to move away from relationality to the individual is to take us back to the problematic path of liberal humanism in which the Western, white model of the individual becomes the measure—of the human and of the dog. I very much concur with Motamedi-Fraser that the 'dog product'—that is, living together—requires scrutiny. She is right to suggest that the behavioural problems so many dogs are said to possess are likely to be examples of resistance to, and judgement of, the dog–human bond which is usually considered to define dogs.

Haraway's favourite dog trainer and thinker, Vicki Hearne, would concur to some degree, but as she suggests in *Adam's Task*, 'We are in charge already, like it or not' (Hearne 2000, 48–49). For Hearne, training is more than the enactment of domination in order to produce the dog–human bond and acceptable dog behaviour, it is productive of a relationship and a two-way process. In the notion of 'muddied living', I hope to have captured the two-way relationality of power, despite species inequality, in dog–human relations. Erica Fudge concludes her book *Pets* with the observation that

> It is not only animals that need to be trained to co-habit. Living as a human requires careful training; it is premised on the monitoring of boundaries, and needs us to be involved in compassionate contemplation. (2008, 110)

For Haraway, it is through companionship that positive relationality is forged, and this requires the constant exercise of Fudge's compassionate contemplation because:

> Just who is at home must be permanently in question. The recognition that one cannot know the other or the self, but must ask in respect for all of time who and what are emerging in relationship is key. That is so for all true lovers, of whatever species. (2003, 50)

For Haraway, agility training is a way of forging ethical relations between species in muddied everyday encounters. Sociologist Leslie Irvine reflects on 'being beginners' at an agility class with Skipper, her dog. Skipper eventually seems to enjoy the class, but in making the course less daunting for him, Irvine finds herself crawling through tunnels and over ground with recently melted snow. After the class, walking back to her car, she reflects on

> being a week away from my forty-second birthday, playing in the mud with my dog. I do not think I have ever been that muddy before, surely not in my adult life. Skipper was muddy too. He stayed at my side and seemed to know we had done something good together . . . I was happy. Everything about me became profoundly beautiful. As the sun set behind the mountains, it streaked the sky with pink. I saw a red-tailed hawk roosting in a tree. The air smelled faintly of wood smoke. Ordinarily, I am something of a worrier, but at that moment I did not have a care in the world. The two of us who do not share a language nevertheless communicated, cooperated, and had fun. (2004, 171)

In walking, in sharing home, in making kin, in caring and being cared for, we need constantly ask, 'Who are you?', 'Who am I?' and 'What does it mean to be in relation?'. We also need to be more dog. Dogs are intensely social and exploratory beings, they are not self-conscious, they do not pretend (except in play) as far as I am aware—their emotions are unfiltered, certainly in comparison to people like me. As Jeffrey Masson puts it:

> Nobody has ever seen a sad dog pretending to be happy or a happy dog pretending to be sad. When a dog is sad or happy, that feeling occupies its whole being; the dog becomes pure happiness or pure sadness. (Masson 1998, 47)

I think it fair to say that Irvine, in playing in the mud with Skipper, had become more dog. 'I was happy', she says, 'at that moment I did not have a care in the world'. At that moment, Irvine is *in the moment*.

Anyone who has played with dogs or simply walked with dogs will know that dogs live in the moment—chasing a rabbit or a ball, watching a squirrel, greeting or playing with a friend. It is not that people do not do this, but that this is relatively rare for adult humans, raised in Western societies to Western norms, to exhibit the sensate manner of being-in-the-world that dogs present. Living an embodied life in the moment, being less guarded, more emotionally responsive and inhabiting a different subjectivity where we really notice what is happening around us is something from which many people may benefit. Hanging around with dogs who are (at least to some degree) unleashed, and enabled to behave like dogs, is probably a more enjoyable route to this than an online mindfulness workshop. *Being more dog*, however, requires making subjective space to share—both with animal companions and beyond. This implies a corollary—the need to *be less human*.

BE LESS HUMAN

Together with Steve Hobden (Cudworth and Hobden 2018), I have argued that the context of environmental crises, or, more precisely, the existential threat posed by global warming and mass extinction, compels a radical reconsideration of what it means to be human on a damaged planet. From being kin with dogs, can we extend our imagination and think about our interactions and co-dependencies with multiple beings and things? Can sharing heterotopic space lead to a prefigurative politics?

It is important to remember that the Western idea of the human is an historical product. The work of Norbert Elias is useful here, suggesting that the processes of modernisation in Europe were about civilising the human animal, about uplift from the animalic. Elias considers how post-medieval European mores on matters of sexuality, bodily functions, violence, table manners and speech were transformed over time through the internalisation of self-restraint and 'expanding the threshold of repugnance' (2000, 71). Increased repression of violent behaviour and restrictions on the freedom to engage in belching, farting, scratching, copulating and so on in public is about the taming of the human animal. The civilised European subject was to expunge their animality via self-discipline. Furthermore,

> European encounters with unfamiliar creatures, human and otherwise in colonized lands, led to the racialization of this animalized notion of the 'uncivilized' and to the domination (through policies of extinction and replacement, or control) of the human, non-human animals and plant life that offended the 'temperate' culture of the colonizing powers. The impact on the non-human lifeworld of European modernity has been profound. (Cudworth and Hobden 2018, 215)

Ecofeminist work has long made such connections, both in terms of historical cases for the gendering of European development and the interlinked marginalisation's of nonhuman animals, nature, women, colonised and indigenous peoples (see Cudworth 2005). Careful to note that no claims are universally true of all indigenous peoples, Val Plumwood (2002) draws on particular examples (primarily from Aboriginal Australian and Native American cultures) to argue that interspecies communicative ethics is required in undoing the violences of Western imperial practices, both in terms of their historical legacies and contesting current forms. Such cultures, she argues, are grounded in an ethics of belonging and community (rather than conquest and private property) and on flourishing (rather than wealth), which assume embodied material interdependencies (Plumwood 2002, 223). She suggests that we need to end quests for transcendence and immortality and understand that we are embodied creatures, amongst others, grounded in place and sustained by social-natural systems. Rather differently, yet in the same vein, John Gray (2013, 164 and 208) argues that it would help our relations with other peoples and certainly with the planet and its myriad species, if we could stop seeking redemption from ourselves through some kind of transcendence and simply appreciate our being in the world and our intimate connections to it.

In *The Emancipatory Project of Posthumanism*, Steve and I took up Anat Pick's use of 'creaturely' as a term that transcends the human/animal binary while emphasising the 'material, temporal and vulnerable' properties of creatureliness (Pick 2011, 5). There are two implications of reading through a 'creaturely prism' for Pick. First, recognising creatureliness means that we might contract the human in order to make room for animals in 'our world of subjectivity' (2011, 6). Second, creatureliness grounds our political standpoint in 'lives' rather than 'rights' (2011, 11). We have advocated a 'creaturely politics' which involves bodied experience in collective spaces and where we think better ways of getting along together might be forged. This requires not a rejection of the human but a shrinking of the Western frame of being human, perhaps through an appreciation of how non-imperial humans have tried to be in the world and a reinvention of new possibilities for sharing place.

A politics of being in the world, of interspecies attachments, of appreciation of relationality, of recognition of entanglement has been suggested by the research underpinning this book. As such, this book is framed by new materialism. It is also heavily influenced by ecofeminist concerns and the perspectives of feminist animal studies. There is, it must be acknowledged, a contradiction here, and one which *The Emancipatory Project* tried to tackle head on—in a world of entanglement and relational complexity, what does this mean for relationships of dominatory power? and how do we tackle the marginalisation, exploitation and oppression of animals? As Eva Giraud

(2019) asks, once we recognise the entanglements of multispecies worlds, what next? Recognition of entanglement does not, as Giraud rightly points out, necessarily lead to less anthropocentric actions. As chapter 4 of this book, in particular, has shown, anthroparchal relations of the exploitation of nonhuman animals remain uncontested in multispecies homes. If we strive to 'be more dog' and live in the moment, if we appreciate the range of our attachment sites (to follow Latour and Haraway), does this preclude action to improve the lives of animals? The final section of this book will now consider some partial and no doubt inadequate answers to how we might live better with dog companions.

LIVING BETTER IN ZOOPOLIS

The lack of compassion and lack of contemplation that characterise relationships between people and dogs in the relatively wealthy societies of the Global North is stark and troubling. The mass violence of forced breeding for the pet industry and the disposability of dogs as 'mere pets', as simply property, is staggering, as we saw at the start of this book. The use of other mammals, of birds and fishes, as food for animal companions is also hugely consequential for the lives of those animals which become dog food, and in terms of the wider environmental implications. If we are not to treat animals, particularly those we live with, as companions, as things, what does this mean?

For Irvine (2004), recognising the selfhood of other creatures and recognising the value of animal lives means that we cannot breed animals and keep them for our pleasure. For Motamedi-Fraser (2024) uncomfortable questions are raised by understanding the oppressive qualities of 'the bond' which is constituted as the human-dog relation. While I understand this, I wonder what the implications are for dogs—re-wilding? A feral existence? The extinction of the dog as we now know it? As a critical posthumanist, I consider it is vital to understand our living world as co-constituted, alongside a critique of oppressive power. Creaturely politics, foregrounding the shared embodied vulnerability of species (animal and beyond!) necessitates an opposition to the instrumentalisation, enslavement and destruction of nonhuman beings (see Cudworth and Hobden 2018, 90–116). This does not mean liberating animals in the sense of withdrawing from them, abolishing all use of nonhuman animals and not engaging with them (MacCormack 2020). This is impossible and illusory—as Sue Donaldson and Will Kymlica (2011)[1] make clear. In their *Zoopolis*, we are always in encounter with other animals, wild, domesticate or otherwise, and human separatism is a false utopia. Rather, humans have always been entangled, intra-dependent with other animals, and every individual is a multispecies crowd.

The task of kin-making with dogs is to stay with the trouble of entanglements of love and of power. What might come after entanglement (to follow Giraud's provocation) for people living with dogs? Kin should not be a property relation, and although I have referred to dog owners throughout this book, this needs transformation. I have never used the term guardian as it does not reflect the current reality for dogs, who are owned. However, it should reflect a reality, and the legal status of dogs[2] needs to change. This does not imply, in my view, that people should no longer live with dogs but it does mean that dogs should not be commercially bred or bought and sold, treated with violence or euthanised at the whim of the humans they live with. While the idea of guardians of adult dogs does have a whiff of paternalism, it is preferable to ownership and implies responsibility or response-ability and caring obligation. When I wrote my will, a solicitor was amazed that my partner and I made financial (and other) provision for any nonhuman animals that may live with us at the time of our deaths. It was our responsibility. Guardians would also need to think about what might be appropriate and most ethical for those creatures for whom they were responsible, for example, with respect to neutering and/or contraception.

Being gregarious, dogs need to socialise and exercise. Responsibility means that for many dogs in Western countries, free roaming is not possible. Yet we need to keep the dogs taking themselves off for a sunbathe on Walthamstow Marshes in our mind's eye, and follow the advice of Bekoff and Pierce on unleashing dogs as far as possible. In countries like the UK, public open space is in urgent need of protection, and space in which dogs can be dogs needs fighting for. The human workplace has a significantly negative impact—on both people and dogs—and flexible human working hours and homeworking could help greatly in tackling the problem of separation anxiety, as could affordable, appropriate and regulated commercial day-care provision, or, more desirable in my view, mutual aid in dog sitting and sharing within kin groups and communities. As long as they are not left for overly long periods, or excessively restricted, the home can be a place of relative freedom for dogs, but the possibilities for enhancing that freedom should be constantly in question for dog guardians. Prejudice against people with dogs in crisis needs to be tackled urgently, for example, in projects to address homelessness, generally, and in the context of domestic violence and abuse (Cudworth and Turgoose 2023). The use of animals as food, for humans or for dogs, is of urgent need of transformation and is a wider structural issue. Suffice it to say here that the feeding of dog companions needs some serious thought—the transformation of animals currently used in agriculture, from property to creatures for whom people have response-ability, would be revolutionary, indeed, and is necessary politically and ethically as part of wider endeavours in the context of climate change and extinction. It would not, as those such as Porcher fear, mean a 'rupture' of our relations with goats, or sheep or cattle. It would mean an end to forced reproduction, killing and commercial exchange of these animals, which would make room for such creatures in human worlds of subjectivity. As this, sadly, is not something I will personally witness in my lifetime, I will say for now that opening veterinary training and practice, and the companion animal food industry, to developing and promoting plant-based diets for dogs would be a small but necessary and achievable step towards living less wrongly. An ethics of care and compassion in our relations with companion dogs is bound up with wider issues of kin-making, for sure.

The social, as Bob Carter and Nickie Charles (2018) emphasise, *is* animal, and we cannot get away from that. The huge numbers of captive dogs in wealthier countries, occupying homes and human hearts, muddying beds and relationships, deserve full recognition as homemakers, co-walkers, carers and kin. They deserve change in the circumstances, the boundaries, within which they are able to act. It matters what relations we make and which we break, and it matters how, what and who we struggle for. Entangled lives in the mongrel domestic are always emerging and can shift and change. For those readers who live or who have lived with dog companions in particular, let us try and secure change which is more sustainable and more nourishing for dogs and their people. In that process, perhaps we can try and be a little less human, and be a bit more dog.

Notes

INTRODUCTION

1. I refer to people living with dogs as companions in their homes as owners. My use of the term has been criticised in work I have submitted to journals, with reviewers objecting that this objectifies dogs, reflects humancentrism and should be replaced with the term guardian. I resist this because the legal conditions of being a dog in the UK is that dogs are property (albeit with certain protections as sentient creatures) and they have human owners. People living with dogs also refer to themselves as 'owners', never, in my experience, as guardians. Finally, guardian, while being associated with legal responsibility also has the tinge of paternalism and surrogate parenthood that will be questioned later in this book.

2. Conversation with a dog walker working for the Canal and River Trust, a charitable organisation that manages the more than two thousand miles of historic waterways across England and Wales.

3. The word murder in English refers to the premeditated and unlawful killing of one human being by another.

4. Jessica Murray, 'American XL Bully Dog Will Be Banned, Says Rishi Sunak', *The Guardian*, 15 September 2023, https://www.theguardian.com/lifeandstyle/2023/sep/15/american-xl-bully-dog-will-be-banned-says-sunak.

5. *The Guardian* newspaper has reported that animals are being taken to vets in order to be euthanised, leading to objections by veterinarians and charities voicing concern about the impact of potential destruction of dogs in their care on staff: Jessica Murray, 'Britons Have Started Handing In XL Bullies to Be Put Down, MPs Told', *The Guardian*, 18 October 2023, https://www.theguardian.com/uk-news/2023/oct/18/britons-have-started-handing-in-xl-bullies-to-be-put-down-mps-told.

6. The only references I can now find in the press after so long are in the *Scottish Herald*, 'Terror as Rottweilers Attack Pupils', 14 February 1990, https://www.heraldscotland.com/news/11974280.terror-as-rottweilers-attack-pupils/; and 'School Attack Dogs to Be Destroyed', 10 May 1990, https://www.heraldscotland.com/news/11944251.school-attack-dogs-to-be-destroyed/; and it merits a mention in the 'history' of the school in its Wikipedia entry, s.v. 'Bishop Stopford's School', last modified 28 October 2023, https://en.wikipedia.org/wiki/Bishop_Stopford%27s_School.

CHAPTER 1

1. The times of these observations were similar in the mornings but varied in the afternoon/evening.

2. I did not consider it necessary to repeat a period of observation in the second site because my daily dog walking experiences did not indicate that I would gain different insights into human-dog relations than in the London site.

3. Thus interviewees in both sites were previously known to me from dog walking encounters. No one I asked declined to participate, although it was far more difficult to pin down potential male participants. All the women who agreed to be interviewed were interviewed, but I failed to interview five of the men who had agreed to participate within the timeframe.

4. A contributing factor to this may have been the closure of Walthamstow dog racing track in 2008, at the time, the largest greyhound racing venue in Britain. Local kennels rehomed large numbers of dogs, many locally.

5. I knew the occupation of my many interviewees prior to interview, if not I asked this. The majority of interviewees were in paid employment. A minority were retired, were at home with young children or did not formally work.

6. The personality of Kevin and Ruby was the only factor of significance here.

7. A lurcher is a mixture of some kind of sighthound (for example, a deerhound, greyhound, saluki or whippet) with a terrier or a herding dog (such as a border collie). Gypsy, for example, was the offspring of a whippet father and a mother who was greyhound, with some deerhound and border collie. The term is also used to describe longdogs, which are mixtures of different kinds of sighthounds, so Ruby, for example, is a mixture of whippet, greyhound and saluki.

8. All the human participants in this study were given an information sheet and signed a consent form approved. While the University Ethics Committee, which approved the project, raised issues about human participants which I had to respond to, no mention was made of the dogs who were clearly present in the interview process. At the time, wanting to get on with the research, I did not raise the issue of nonhuman participation, but on reflection, this kind of research in the social sciences raises questions for current institutional procedures.

CHAPTER 2

1. An organisation with royal patronage since its inception, and as of 2023, with a royal prefix—the Royal Kennel Club, see https://www.thekennelclub.org.uk/media-centre/2023/april/royal-prefix-on-150th-anniversary/.

2. Meaning off-leash.

3. See www.leevalleypark.org.uk.

4. This was met with various degrees of tolerance, usually depending on the amount of litter left, or the extent to which dogs might be injured as a result of detritus (broken glass, discarded or lost drugs and the other bane of dog owners, drug-laced human

poop and the odd allure it holds for dogs). It seems the popularity of the marshes as a venue is retained as I write this, to the chagrin of those living close by (Kelly 2022).

5. As we saw in chapter 1, this could also apply to dogs as research assistants.

6. For example, see 'Fabsternation Expresses Consternation at State of Leyton March', Save Lea Marshes (SLM) blog, posted 22 December 2012. http://saveleytonmarsh.wordpress.com/2012/12/22/fabsternation-expresses-consternation-at-state-of-leyton-marsh/.

CHAPTER 3

1. See *Cleanipedia*, 'How to Get Rid of Dog Smell in Your Home', last updated 23 August 2022, https://www.cleanipedia.com/gb/family/get-rid-dog-smell-home.html.

2. *wikiHow* and Emily Fazio, 'How to Eliminate Dog Smell', 26 October 2022, https://www.wikihow.com/Eliminate-Dog-Smell#:~:text=Simply%20add%20two%20tablespoons%20of,clean%20with%20a%20damp%20cloth.

3. See Danielle Fisher, 'Five Pet Problems to Fix Before You Sell Your House', *HowStuffWorks*, accessed 21 March 2023, https://home.howstuffworks.com/real-estate/selling-home/5-pet-problems-to-fix-before-sell.html.

4. See Cedars, 'Dirty Dogs, Clean House?', *Horse & Hound* forum thread post, posted 19 May 2009, https://forums.horseandhound.co.uk/threads/dirty-dogs-clean-house.637373/.

5. See Jennifer Ebert, 'Seven Ways to Make Your Home More Dog-Friendly', *Ideal Home*, 7 November 2017, https://www.idealhome.co.uk/news/home-dog-friendly-187491.

6. This not to suggest Ingold has rejected human exceptionalism. When it comes to homemaking, only human beings are 'the authors of their own designs' (2012, 175).

7. As not all may know, tomato ketchup is the solution to the lingering and pungent smell of fox shit on the coat of a dog. For those readers who might benefit, the instructions are as follows: rinse off the worst of the shit, apply ketchup and leave for a minute, wash off and shampoo dog/affected area. Referring to fox shit—a favoured perfume of many dogs—as 'Channel No. 5' is quite common in England.

CHAPTER 4

1. This term is used within UK slaughterhouses to describe the skip where the workers throw the parts of animals that will be turned into processed meat products such as burgers and sausages.

2. Various kinds of tests might be used, from the more benign taste trails to toxicity testing whereby dogs will be made ill, often critically so, and in some forms of testing (such as 'lethal dose'), will die.

3. Beagles are a popular breed for use in animal testing as they are gentle dogs who tend to be kind-natured towards humans and other dogs.
4. A relatively small 'sister' company of a large US firm which has been a target of protest.
5. The logo and quote can be found at https://www.epicicons.com/work, under 'food'.
6. See https://www.jrpetproducts.com/collections/primal-range-with-hair.
7. The General Data Protection Regulation of the European Union Economic Area which governs how personal data can be used, processed and stored.
8. 'Quid' is an English slang expression for the British pound sterling (GBP), it may stem from the Latin phrase *quid pro quo*, meaning something which you exchange for something else.
9. Unsurprisingly, Pure changed their advertising from 2020 to 2022. While the 2020 advertisements showed the food, those in 2022 did not. As the food which was marketed as fresh and delicious was shown to be dried powder which reconstituted into an unappetising looking slurry, perhaps this was not surprising!

CHAPTER 5

1. The countries were numerous, including Austria, Belgium, Bulgaria, Croatia, the Czech Republic, Denmark, Estonia, Finland, France, Hungary, Ireland, India, Italy, Lithuania, Malta, Moldova, the Netherlands, Poland, Portugal, Slovenia and Slovakia. Some countries provided a high degree of ease of entry and support, such as Germany, Greece, Latvia and Romania, while others, such as Norway and the UK, made some but far more limited accommodation. Details of each countries policies can be found on the International Fund for Animal Welfare (IFAW) webpages, https://www.ifaw.org/news/resources-flee-ukraine-pets.
2. The population questions do raise questions for me, but Haraway has faith in a more humbled future human and is sensitive to profound inequalities and the way these impact the violences of making live and letting die, unlike the 'ahuman' position of Patricia MacCormack (2020) where embracing the actual death of humanity is required for the thriving of other species. A dark ecology, indeed.
3. Country is a belief system and sense of identity and belonging for Aboriginal and Torres Strait Islander peoples in Australia. It refers to place but is more than land, including the whole collection of animals, plants, people, ancestors, creation spirits, waterways and so on that constitute place.
4. Interestingly, this woman came from a farming family which raised sheep.
5. A lush, in this context, is slang for a habitual heavy drinker, a drunk.
6. The original marketing name for diazepam, used to treat anxiety and depression. Its use in the 1960s in countries such as the United States and the United Kingdom became notorious, as it was popularly described as 'mother's little helper' and considered a cure for unhappy housewives despite its addictive properties. See, for example, Paul Schleis, 'The Drug that Became Prescription for Trouble', *The Washington Post*, 8 May 2005, https://www.washingtonpost.com/archive/

lifestyle/2005/05/08/the-drug-that-became-prescription-for-trouble/2ebf91cc-a16c-4b38-881c-0c4112e8aa0c/.

7. Staffordshire bull terriers.

CHAPTER 6

1. Thanks to Matthew Cole for alerting me to this paper during a heated discussion of the work of Richard Scarry!

2. Some of the data David Redmalm analyses is from television, specifically *The Simple Life* show, starring Paris Hilton and her Chihuahua, Tinkerbell. I would like to smuggle in congratulations to him here for such a great title for a journal article. It makes me smile every time I see it.

3. Such as the RSPCA campaign #DogKind, https://www.rspca.org.uk/documents/1494939/7712578/%23DogKind+Being+left+alone+%28PDF+1MB%29.pdf/d5e12b32-19a0-be1d-6280-619cdb322d1d?t=1588152649339.

4. It is important to note that the data for this project was collected prior to the emergence of the COVID-19 pandemic in the UK in 2020.

CONCLUSION

1. Patricia MacCormack's controversial thesis is that human beings abolish their use and exploitation of animals and actively promote the abolition of the human by not reproducing.

2. Other animals also, which would be a radical transformation, indeed, and would necessitate significant changes in human practice.

References

Aaron, Jane, and Syliva Walby. 1991. *Out of the Margins: Women's Studies in the Nineties*. London: Falmer.

Adams, Carol J. 1995. 'Woman Battering and Harm to Animals'. In *Animals and Women: Feminist Theoretical Explorations*, edited by Carol J. Adams and Josephine Donnovan, 55–84. Durham, NC: Duke University Press.

Alger, Janet M., and Steven F. Alger. 1999. 'Cat Culture, Human Culture: An Ethnographic Study of a Cat Shelter'. *Society and Animals* 7, no. 3: 199–218.

Alger, Janet M., and Steven F. Alger. 2003a. 'Drawing the Line Between Humans and Animals: An Examination of Introductory Sociology Textbooks'. *International Journal of Sociology and Social Policy* 23, no. 3: 69–93. https://doi.org/10.1108/01443330310790264.

Alger, Janet M., and Steven F. Alger. 2003b. *Cat Culture: The Social World of a Cat Shelter*. Philadelphia, PA: Temple University Press.

Allcock, Toby, and Madeleine L. H. Campbell. 2021. 'The UK Dangerous Dogs Act: Improved, but Legally and Ethically Flawed'. *The Veterinary Record* 189, no. 3: 85–86. https://pubmed.ncbi.nlm.nih.gov/33893662/.

Anderson, Bridget. 2000. *Doing the Dirty Work? The Global Politics of Domestic Labour*. London: Zed.

Anderson, Gillian., Joseph G. Moore and Laura Suski. 2016. 'Introduction: For a Sociology of Home in Canada'. In *Sociology of Home: Belonging, Community and Place in the Canadian Context*, edited by Gillian Anderson, Joseph G. Moore and Laura Suski, 1–12. Toronto: Canadian Scholars Press.

Arcari, Paula. 2020. *Making Sense of Food Animals: A Critical Exploration of the Persistence of Meat.* London: Palgrave Macmillan.

Arendt, Hannah. 1998. *The Human Condition*, second edition. Chicago: Chicago University Press.

Arluke, Arnold. 2003. 'Ethnozoology and the Future of Sociology'. *International Journal of Sociology and Social Policy* 23, no. 3: 26–45.

Arluke, Arnold. 2004. 'A Sociology of Sociological Animal Studies'. *Society and Animals* 10, no. 4: 369–74.

Arluke, Arnold, and Clinton Sanders. 1996. *Regarding Animals*. Philadelphia: Temple University Press.

Ascione, Frank R., Claudia V. Weber and David S. Wood. 1997. 'The Abuse of Animals and Domestic Violence: A National Survey of Shelters for Women who are Battered'. *Society and Animals* 5, no. 3: 205–18. https://doi.org/10.1163/156853097X00132.

Ashall, Vanessa, and Pru Hobson-West. 2017. '"Doing Good by Proxy": Human-Animal Kinship and the "Donation" of Canine Blood'. *Sociology of Health and Illness* 39, no. 6: 908–22.

Barkham, Patrick, and Simon Murphy. 2012. 'Bark but No Bite: Dangerous Dogs Act in Spotlight as Attacks Rise'. *The Guardian*, 18 January 2012. https://www.theguardian.com/uk/2012/jan/18/dangerous-dogs-act-in-spotlight.

Barnes, Brookes. 2014. 'Lassie-as-Salesdog: One More Trip to the Well?' *New York Times*, 8 August 2014. http://www.nytimes.com/2014/08/17/business/media/lassie-as-salesdog-one-more-trip-to-the-well.html?_r=0.

Barrett, Michèle. 1980. *Women's Oppression Today: The Marxist/Feminist Encounter*. London: Verso.

Bates, Charlotte, and Alex Rhys-Taylor. 2017. 'Finding Our Feet'. In *Walking Through Social Research*, edited by Charlotte Bates and Alex Rhys-Taylor, 1–11. Abingdon: Routledge.

Bauman, Adrian E., J. Russell Schroeder, Susan E. Furber and Annette J. Dobson. 2001. 'The Epidemiology of Dog Walking: An Unmet Need for Human and Canine Health'. *Medical Journal of Australia* 175, nos. 11–12: 632–34. https://doi.org/10.5694/j.1326-5377.2001.tb143757.x.

BBC. 2017. 'Italian Woman Granted Sick Pay to Look after Ill Dog'. 12 October 2017. https://www.bbc.co.uk/news/world-europe-41594224.

Bear, Christopher, Katy Wilkinson and Lewis Holloway. 2017. 'Visualising Human- Animal-Technology Relations: Fieldnotes, Still Photography and Digital Video on the Robotic Dairy Farm'. *Society and Animals* 25, no. 3: 225–56. https://doi.org/10.1163/15685306-12341405.

Beck, Alan, and Aron Katcher. 1983. *Between Pets and People.* New York: Putnam's.

Beck, Ulrich, and Elisabeth Beck-Gernsheim. 1995. *The Normal Chaos of Love.* Cambridge: Polity Press

Bekoff, Marc. 2002. *Minding Animals: Awareness, Emotion and Heart.* New York: Oxford University Press.

Bekoff, Marc. 2007. *The Emotional Lives of Animals*. Novato, CA: New World Library.

Bekoff, Marc, and Jessica Pierce. 2019. *Unleashing Your Dog: A Field Guide to Giving Your Canine Companion the Best Life Possible.* Novato, CA: New World Library.

Belshaw, Zoe, Lucy Asher and Rachel Dean. 2019. '"You Can be Blind Because of Loving Them So Much": The Impact on Owners in the United Kingdom of Living with a Dog with Osteoarthritis'. *BMC Veterinary Research* 16, no. 1: 85. https://doi.org/10.21203/rs.2.19056/v1.

Benjamin, Ruha. 2018. 'Black After Lives Matter: Cultivating Kinfulness as Reproductive Justice'. In *Making Kin Not Population*, edited by Adele Clarke and Donna Haraway, 41–66, Chicago, IL: Prickly Paradigm Press.

Berger, Lawrence M., and Marcia J. Carson. 2020. 'Family Policy and Complex Contemporary Families: A Decade in Review and Implications for the Next Decade of Research and Policy Practice'. *Journal of Marriage and Family* 82, no. 1: 478–507.

Best, Steve. 2010. 'Animal Liberation, Human Liberation and the Future of the Left'. http://www.drstevebest.org/RethinkingRevolution.htm.

Best, Steve. 2014. *The Politics of Total Liberation: Revolution for the 21st Century*. New York: Palgrave.

Bird Rose, Deborah. 1992. *Dingo Makes Us Human: Life and Land in Aboriginal Australian Culture.* Cambridge: Cambridge University Press.

Bird Rose, Deborah. 2011. *Wild Dog Dreaming: Love and Extinction*. Charlottesville, VA: University of Virginia Press.

Birke, Lynda. 2014. 'Listening to Voices: On the Pleasures and Problems of Studying Human-Animal Relationships'. In *The Rise of Critical Animal Studies: From the Margins to the Centre*, edited by Nik Taylor and Richard Twine, 71–87 Abingdon: Routledge.

Birke, Lynda, and Jo Hockenhull. 2012. 'On Investigating Human-Animal Bonds: Realities, Relatings, Research'. In *Crossing Boundaries. Investigating Human-Animal Relationships*, edited by Lynda Birke and Jo Hockenhull, 1 –15. Leiden: Brill.

Blouin, David. 2013. 'Are Dogs Children, Companions, or Just Animals? Understanding Variations in People's Orientations Toward Animals'. *Anthrozoös* 26, no. 2: 279–94. https://doi.org/10.2752/175303713X13636846944402.

Bowlby, Sophie, Susan Gregory and Linda McKie. 1997. ''Doing home': Patriarchy, caring and space', *Womens Studies International Forum* 20, no. 3: 343–50.

Boyd, Caitlin, Susan Jarvis, Paul D. McGreevy, Sarah Heath, D. B. Church, David C. Brodbelt and D. G. O'Neill. 2018. 'Mortality Resulting from Undesirable Behaviours in Dogs Aged Under Three Years Attending Primary-Care Veterinary Practices in England'. *Animal Welfare* 27, no. 3: 251–62. https://doi.org/10.7120/09627286.27.3.251.

Bradshaw, John. 2012. *In Defence of Dogs: Why Dogs Need Our Understanding*. London: Penguin.

Braidotti, Rosi. 2013. *The Posthuman*. Cambridge, United Kingdom: Polity.

Brown, Wendy, Barbara Vanselow, Andrew J. Redman and John R. Pluske. 2009. 'An Experimental Meat-free Diet Maintained Haematological Characteristics in Sprint-racing Sled Dogs'. *British Journal of Nutrition* 102, no. 9: 1318–23. doi:10.1017/S0007114509389254.

Budiansky, Stephen. 2001. *The Truth About Dogs: The Ancestry, Social Conventions, Mental Habits and Moral Fibre of Canis Familiaris.* London: Weidenfeld and Nicolson.

Butchers. 2021. 'Nourishing Food for Dogs' 2021 TV Advert. Accessed 31 March 2023. https://www.youtube.com/watch?v=lGkNZYcMH4c.

Butternut Box. 2022. TV Advert. Accessed 31 March 2023 https://www.youtube.com/watch?v=Yj268G2V5-A.

Butternut Box. 2023. https://butternutbox.com/our_products.

Callicott, John B. 1980. 'Animal Liberation: A Triangular Affair'. In *In Defense of the Land: Essays in Environmental Philosophy*, edited by John Baird Callicott, 15–38, Albany: SUNY.

Carpiano, Richard M. 2009. 'Come Take a Walk with Me: The 'Go-along' Interview as a Novel Method for Studying the Implications of Place for Health and Well-being'. *Health and Place* 15, no. 1: 263–72.

Carter, Bob, and Nickie Charles. 2018. 'The Animal Challenge to Sociology'. *European Journal of Social Theory* 21, no. 1: 79–97.

Cedars, 'Dirty Dogs, Clean House?', *Horse & Hound,* forum thread post, posted 19 May 2009, https://forums.horseandhound.co.uk/threads/dirty-dogs-clean-house.637373/.

Chadderton, Charlotte. 2013. 'Secondary Schools Under Surveillance: Young People "As" Risk in the UK: An Exploration of the Neoliberal Shift from Compassion to Repression'. In *Technology, Society and Inequality: New Horizons and Contested Futures*, edited by Erika Cudworth, Peter Senker and Kathy Walker, 139–54. New York and London: Peter Lang.

Challenger, Melanie. 2021. *How to Be Animal: A New History of What It Means to Be Human.* Edinburgh: Canongate.

Chambers, Deborah, and Pablo Gracia. 2021. *A Sociology of Family Life: Change and Diversity in Intimate Relations*, second edition. Cambridge: Polity.

Chapman, Tony, and Jenny Hockey. 1999. 'The Ideal Home as It Is Imagined and as It Is Lived'. In *Ideal Homes? Social Change and Domestic Life*, edited by Tony Chapman and Jenny Hockey, 1–14. London: Routledge.

Charles, Nickie. 2014. '"Animals Just Love You as You Are": Experiencing Kinship Across the Species Barrier'. *Sociology* 48, no. 4: 715–30.

Charles, Nickie. 2016. 'Post-Human Families? Dog-Human Relations in the Domestic Sphere'. *Sociological Research Online* 21, no. 3: 83–94.

Charles, Nickie, and Charlotte Aull Davies. 2008. 'My Family and Other Animals: Pets as Kin'. *Sociological Research Online* 13, no. 5. http://www.socresonline.org.uk,13/5/4.html.

Charles, Nickie, and Carol Wolkowitz. 2019. 'Bringing Dogs onto Campus: Inclusions and Exclusions of Animal Bodies in Organizations'. *Gender, Work and Organization* 26, no. 3: 303–21.

Charles, Nickie, and Carol Wolkowitz. 2023. "Basically He's a Pet, Not a Working Dog': Theorising What Therapy Dogs Do in the Workplace'. *Work, Employment and Society* 0, no. 0: 1–22. https://doi.org/10.1177/09500170231161495.

Charles, Nickie, Rebekah Fox, Harriet Smith and Mara Miele. 2021. '"Fulfilling Your Dog's Potential": Changing Dimensions of Power in Training Cultures in the UK'. *Animal Studies Journal* 10, no. 2: 169–200.

Christiansen, Stine B., Annemarie T. Kristensen, Peter Sandøe and Jesper Lassen. 2013. 'Looking After Chronically Ill Dogs: Impacts on the Caregiver's Life'. *Anthrozoös* 26, no. 4: 519–33. https://doi.org/10.2752/175303713X13795775536174.

Cimarelli, Guilia, Borbála Turcsán, Friederike Range and Zófia Virányi. 2017. 'The Other End of the Leash: An Experimental Test to Analyze How Owners Interact with their Pets'. *Journal of Visualized Experiments* 128: e56233. doi: 10.3791/56233 28.

Clarke, Adele E. 2018. 'Introducing Making Kin Not Population'. In *Making Kin Not Population*, edited by Adele E. Clarke and Donna J. Haraway, 1–40. Chicago, IL: Prickly Paradigm Press.

Clarke, Adele E., and Donna J. Haraway, eds. 2018. *Making Kin not Population*. Chicago, IL: Prickly Paradigm Press.

Clarke, Caroline A., and David Knights. 2018. 'Practice Makes Perfect? Skillful Performances in Veterinary Work'. *Human Relations* 71, no. 10: 1395–421.

Clark, John L. 2014. 'Labourers or Lab Tools? Rethinking the Role of Lab Animals'. In *The Rise of Critical Animal Studies: From the Margins to the Centre*, edited by Nik Taylor and Richard Twine, 139–64. Abingdon, UK: Routledge.

Clark, Nigel. 2007. 'Animal Interface: The Generosity of Domestication'. In *Where the Wild Things Are Now: Domestication Reconsidered*, edited by Rebecca Cassidy and Molly Mullin, 49–70. Oxford: Berg.

Cleanipedia, 'How to Get Rid of Dog Smell in Your Home', last updated 23 August 2022, https://www.cleanipedia.com/gb/family/get-rid-dog-smell-home.html.

Clutton-Brock, Juliet. 2016. 'Origins of the Dog: The Archaeological Evidence'. In *The Domestic Dog: Its Evolution, Behavior and Interactions with People*, second edition, edited by James Serpell, 7–21. Cambridge: Cambridge University Press.

Cohen, Susan P. 2002. 'Can Pets Function as Family Members?' *Western Journal of Nursing Research* 24, no. 6: 621–38.

Contagious. 2021. 'Pedigree's GDPR-friendly Campaign Targets Dogs Instead of People'. 6 July. https://www.contagious.com/news-and-views/campaign-of-the-week-pedigree-gdpr-friendly-campaign-targets-dogs.

Cosgrove, Nicole. 2023. '11 UK Animal Shelter Statistics and Facts to Know in 2023'. Accessed 15 February 2023. https://petkeen.com/animal-shelter-statistics-uk/.

Coulter, Kendra. 2016a. *Animals, Work and the Promise of Interspecies Solidarity*. Basingstoke: Palgrave Macmillan.

Coulter, Kendra. 2016b. 'Humane Jobs: A Political Economic Vision for Interspecies Solidarity and Multispecies Wellbeing'. *Politics and Animals* 2, no. 1: 67–77.

Coulter, Kendra. 2016c. 'Beyond Human to Humane: A Multispecies Analysis of Care Work, Its Repression, and Its Potential'. *Studies in Social Justice* 10, no. 2: 199–219.

Coulter, Kendra. 2020. 'Towards Humane Jobs and Work-Lives for Animals'. In *Animal Labour: A New Frontier of Interspecies Justice?*, edited by Charlotte E. Blattner, Kendra Coulter and Will Kymlica, 29–47. Oxford: Oxford University Press.

Cresswell, Tim. 2004. *Place: A Short Introduction*. Oxford: Blackwell.

Cuomo, Chris. 1997. *Feminism and Ecological Communities: An Ethic of Flourishing*. New York: Routledge.

Cudworth, Erika. 2005. *Developing Ecofeminist Theory: The Complexity of Difference*. Basingstoke: Palgrave.

Cudworth, Erika. 2008. '"Most Farmers Prefer Blondes" – Dynamics of Anthroparchy in Animals' Becoming Meat'. *The Journal for Critical Animal Studies* 6, no. 1: 32–45.

Cudworth, Erika. 2010. '"The Recipe for Love"? Continuities and Changes in the Sexual Politics of Meat'. *The Journal for Critical Animal Studies* 8, no. 4: 78–99.

Cudworth, Erika. 2011a. *Social Lives with Other Animals: Tales of Sex, Death and Love*. Basingstoke: Palgrave.

Cudworth, Erika. 2011b. 'Walking the Dog: Explorations and Negotiations of Species Difference'. *Philosophy, Activism, Nature* 8: 14–22.

Cudworth, Erika. 2014. 'Beyond Speciesism: Theorizing the Social Domination of Other Animals'. In *The Rise of Critical Animal Studies: From the Margins to the Centre*, edited by Nik Taylor and Richard Twine, 19–35. London: Routledge.

Cudworth, Erika. 2015. 'Killing Animals: Sociology, Species Relations and Institutionalised Violence'. *The Sociological Review* 63, no. 1: 1–18.

Cudworth, Erika. 2016a. 'A Sociology for Other Animals? Analysis, Advocacy and Intervention'. *International Journal of Sociology and Social Policy* 36, no. 3/4: 242–57.

Cudworth, Erika. 2016b 'On Ambivalence and Resistance: Carnism and Diet in Multispecies Households'. In *Critical Perspectives on Meat Culture*, edited by Annie Potts, 222–42. Leuven: Brill.

Cudworth, Erika. 2017. 'Posthuman Community in the Edgelands'. *Society and Animals* 25, no. 4: 384–403.

Cudworth, Erika. 2018. "Now, Where Were We?' The Highs and Lows of Hunting Data with a Research Pack'. *Journal of Sociology* special issue 'A Sociology of Multi-Species Relations?' 54, no. 4: 488–503.

Cudworth, Erika. 2021. 'Muddied Living: Making Home with Dog Companions'. *International Journal of Sociology and Social Policy* 41, nos. 3–4: 424–39.

Cudworth, Erika. 2022. 'Labors of Love: Work, Labor and Care in Dog-Human Relations'. *Gender, Work and Organization* 29, no. 3: 830–44.

Cudworth, Erika, and Stephen Hobden. 2018. *The Emancipatory Project of Posthumanism*. London: Routledge.

Cudworth, Erika, and Tracey Jensen. 2016. 'Puppy Love? Companion Animals in the Media'. In *Critical Animal and Media Studies: Communication for Nonhuman Animal Advocacy*, edited by Núria Almiron, Matthew C. Cole and Carrie P. Freeman, 185–200. Abingdon: Routledge.

Cudworth, Erika, Ruth McKie and Di Turgoose, eds. 2023. *Feminist Animal Studies: Theories, Practices, Politics*. London: Routledge.

Cudworth, Erika, and Di Turgoose. 2023. 'The Animal Question in Domestic Violence and Abuse – the Intersection Feminism Forgot?' paper presented to the European Conference on Domestic Violence, Reykjavík, Iceland, 11–13 September 2023.

Curtis, Joe. 2013. '450 Sign Petition Against Draconian Dog Control Orders'. *The Guardian*, 27 March. Accessed 21 June 2013. http://www.guardian-series.co.uk/news/10316561.450_sign_petition_against__draconian__dog_control_orders/.

Cutt, Hayley, Billie Giles-Corti and Matthew Knuiman. 2008. 'Encouraging Physical Activity Through Dog Walking: Why Don't Some Owners Walk with Their Dog?' *Preventative Medicine* 46, no. 2: 120–26.

Crave. 2019. 'Hunter' TV advertisement. YouTube video. Accessed 30 March 2023. https://www.youtube.com/watch?v=75b3U3f88kM.

Dalla Costa, Mariarosa, and Selma James. 1973. *The Power of Women and the Subversion of Community*. Bristol: Falling Wall Press.

Dangerous Dogs Act. 1991. http://www.legislation.gov.uk/ukpga/1991/65/contents.

Dashper, Karen. 2019. 'More□than□Human Emotions: Multispecies Emotional Labor in the Tourism Industry'. *Gender, Work and Organization* 27, no. 1: 24–40.

Dearen, Lizzie. 2015. '"I Love My Dog, I Need Her": Syrian Refugee, 17, Walks 300 Miles Carrying Pet to Europe'. *Independent*, 24 September 2015. https://www.independent.co.uk/news/world/europe/i-love-my-dog-i-need-her-syrian-refugee-17-walks-than-300-miles-carrying-pet-to-europe-10511062.html.

De Certeau, Michel. 1984. *The Practice of Everyday Life*. Berkeley, CA: University of California Press.

Deckha, Maneesha. 2012. 'Toward a Postcolonial Posthumanist Feminist Theory: Centralizing Race and Culture in Feminist Work on Nonhuman Animals'. *Hypatia: Journal of Feminist Philosophy* 27, no. 3: 527–45.

Deckha, Maneesha. 2018. 'Humanizing the Non-Human: A Legitimate Way for Animals to Escape Juridical Property Status?' In *Critical Animal Studies: Towards Trans-species Social Justice*, edited by Arturo Matsuoka and John Sorenson, 209–33. Lanham, MD: Rowman & Littlefield.

De Fererr, Martha. 2022. 'The Countries and Organisations Helping Ukraine's Pet Refugees Find a Safe Haven'. *Euronews* 3 March. https://www.euronews.com/green/2022/03/03/the-countries-and-organisations-helping-ukraine-s-pet-refugees-find-a-safe-haven.

De Lauretis, Theresa. 1990. 'Eccentric Subjects: Feminist Theory and Historical Consciousness'. *Feminist Studies* 6, no. 1: 115–50.

Delon, Nicholas. 2020. 'The Meaning of Animal Labour'. In *Animal Labour: A New Frontier of Interspecies Justice?* edited by Charlotte E. Blattner, Kendra Coulter and Will Kymlica, 160–80. Oxford: Oxford University Press.

Dembicki, Diane, and Jennifer Anderson. 1996. 'Pet Ownership May be a Factor in Improved Health of the Elderly'. *Journal of Nutrition for the Elderly* 15, no. 3: 15–31.

Department for Environment, Food and Rural Affairs (Defra). 2009. *Dangerous Dogs Law: Guidance for Enforcers*. London: Department for Environment, Food and Rural Affairs.

Derbyshire, David. 2010. 'Walkies? Fat Chance: One in Five Dog Owners too Lazy to Take Their Pets Out Every Day'. *Daily Mail Online*, 6 August. Accessed 20 March 2023. https://www.dailymail.co.uk/news/article-1300658/Walkics-Fat-chance-One-dog-owners-lazy-pets-day.html.

Derrida, Jacques. 1991. '"Eating Well", or the Calculation of the Subject: An Interview with Jacques Derrida'. By Jean-Luc Nancy. In *Who Comes After the Subject*, edited by Eduardo Cadaver, Peter Connor and Jean-Luc Nancy, 96–119. London: Routledge.

Derrida, Jacques. 2009. *The Beast and the Sovereign*, vol. 1. Edited by Michel Lisse, Marie-Louise Mallet and Ginette Michaud. Translated by Geoffrey. Bennington. Chicago and London: Chicago University Press.

Descola, Philippe. 2013. *Beyond Nature and Culture*. Chicago: University of Chicago Press.

Dickens, Peter. 2004. *Society and Nature: Changing our Environment, Changing Ourselves*. Cambridge: Polity.

Dodd, Sarah A. S., Nick J. Cave, Jennifer L. Adolphe, Anna K. Shoveller and Adroine Verbrugghe. 2022. 'Plant-Based (Vegan) Diets for Pets: A Survey of Pet Owners Attitudes and Feeding Practices'. *Plos One* 17, no. 15: e0268982. https://doi.org/10.1371/journal.pone.0268982.

Donaldson, Jean. 2005. *The Culture Clash*. Berkely, CA: James and Kenneth Publishers.

Donaldson, Sue, and Will Kymlica. 2011. *Zoopolis: A Political Theory of Animal Rights*. Oxford: Oxford University Press.

Donaldson, Sue, and Will Kymlica. 2020. 'Animal Labour in a Post-Work Society'. In *Animal Labour: A New Frontier of Interspecies Justice?* edited by Charlotte E. Blattner, Kendra Coulter and Will Kymlica, 207–28. Oxford: Oxford University Press.

Donovan, Josephine. 2006. 'Feminism and the Treatment of Animals: From Care to Dialogue'. *Signs* 31, no. 2: 305–29.

Donovan, Josephine, and Carol J. Adams. 2007. *The Feminist Care Tradition in Animal Ethics*. New York: Colombia University Press.

Douglas, Mary. 1991. 'The Idea of a Home: A Kind of Space'. *Social Research* 58, no. 1: 287–307.

Dow, Katharine, and Janelle Lamoreaux. 2020. 'Situated Kinmaking and the Population "Problem"'. *Environmental Humanities* 12, no. 2: 475–91.

Drobnič, Sonia. 2011. 'Introduction: Job Quality and Work-life Balance'. In *Work-life Balance in Europe: The Role of Job Quality*, edited by Sonia Drobnič and Ana M. Guillén, 1–14. Basingstoke: Palgrave.

D'Souza, Renée, Alice Hovorka and Lee Niel. 2020. 'Conservation Canines: Exploring Dog Roles, Circumstances and Welfare Status'. In *Animal Labour: A New Frontier of Interspecies Justice?* edited by Charlotte E. Blattner, Kendra Coulter and Will Kymlica, 207–28. Oxford: Oxford University Press.

Dunayer, Joan. 1995. 'Sexist Words, Speciesist Roots?'. In *Animals and Women: Feminist Theoretical Explanations*, edited by Josephine Donovan and Carol J. Adams, 11–31. Durham, NC: Duke University Press.

Ebert, Jennifer. 'Seven Ways to Make Your Home More Dog-Friendly', Ideal Home, 7 November 2017, https://www.idealhome.co.uk/news/home-dog-friendly-187491.

Ehrenreich, Barbara, and Arlie Hochschild, eds. 2002. *Global Women: Nannies, Maids and Sex Workers in the New Economy*. London: Granta.

Elias, Norbert. 2000. *The Civilizing Process: Sociogenetic and Psychogenetic Investigations*. Oxford: Blackwell.

Elwood, Sarah A. 2000. 'Lesbian Living Spaces: Multiple Meanings of Home'. *Journal of Lesbian Studies* 4, no. 1: 11–27.

Escobar, Arturo. 2001. 'Culture Sits in Places: Reflections on Globalism and Subaltern Strategies of Localisation'. *Political Geography* 20, no. 2: 139–74.

Evans, James, and Phil Jones. 2011. 'The Walking Interview: Methodology, Mobility and Place'. *Applied Geography* 31, no. 2: 849–58.

Farley, Paul, and Michael Symmons Roberts. 2012. *Edgelands: Journeys into England's True Wilderness*. London: Vintage.

Fazio, Emily. 'How to Eliminate Dog Smell', *WikiHow,* 26 October 2022, https://www.wikihow.com/Eliminate-Dog-Smell#:~:text=Simply%20add%20two%20tablespoons%20of,clean%20with%20a%20damp%20cloth.

Federici, Sylvia. 2004. *Caliban and the Witch: Women, the Body, and Primitive Accumulation*. New York: Autonomedia.

FEDIAF (European Pet Food Industrial Federation). 2019. 'Facts and figures 2018'. *Annual Report 2019*. Fediaf: Brussels. http://www.fediaf.org.

FEDIAF (European Pet Food Industrial Federation). 2022. 'Facts and Figures 2021: European Dog Population'. *Annual Report 2022*. Fediaf: Brussels. http://www.fediaf.org.

Fife-Cook, Isabel, and Becca Franks. 2021. 'Koi (*Cyprinus rubrofuscus*) Seek Out Tactile Interaction with Humans: General Patterns and Individual Differences'. *Animals* 11, no. 3: 706. doi: 10.3390/ani11030706.

Fisher, Danielle. 'Five Pet Problems to Fix Before You Sell Your House', *HowStuffWorks*, accessed 21 March 2023, https://home.howstuffworks.com/real-estate/selling-home/5-pet-problems-to-fix-before-sell.html.

Fitzgerald, Amy J. 2007. '"They Gave Me a Reason to Live": The Protective Effects of Companion Animals on the Suicidality of Abused Women'. *Humanity & Society* 31, no. 4: 355–78.

Flynn, Clifton P. 2000. 'Battered Women and Their Animal Companions: Symbolic Interaction Between Human and Non-Human Animals'. *Society and Animals* 8, no. 1: 99–127. https://doi.org/10.1163/156853000X00084.

Food and Agriculture Organization of the United Nations. 2002. *World Agriculture: Towards 2015/2030: Summary Report*. ftp://ftp.fao.org/docrep/fao/004/y3557e/y3557e.pdf.

Foucault, Michel. 1975. *Discipline and Punish: The Birth of the Prison*. New York: Random House.

Foucault, Michel. 1986. 'Of Other Spaces'. *Diacritics* 16, no. 1: 22–27.

Foucault, Michel. 1998 [1976]. *The History of Sexuality*, vol. 1, *The Will to Knowledge*. London: Penguin.

Fox, Rebekah. 2006. 'Animal Behaviours, Post-human Lives: Everyday Negotiations of the Animal–Human Divide in Pet-Keeping'. *Social and Cultural Geography* 7, no. 4: 525–37.

Fox, Rebekah. 2017. 'Intimate Cartographies: Creating Place with Companion Animals'. In *Animal Places: Lively Cartographies of Human-animal Relations*, edited by Jacob Bull, Tora Holmberg and Celia Asberg, 85–104. London: Routledge.

Francione, Gary. 2008. *Animals as Persons: Essays on the Abolition of Animal Exploitation*. New York: Columbia University Press.

Franklin, Adrian. 1999. *Animals and Modern Cultures: A Sociology of Human-animal Relations in Modernity.* London: Sage.

Franklin, Adrian. 2006. *Animal Nation.* Sydney: University of New South Wales Press.

Franklin, Adrian. 2007. 'Human-Non-Human Animal Relationships in Australia: An Overview of the First National Survey and Follow Up Case Studies'. *Society and Animals* 15, no. 1: 7–27.

Fraser, Heather, and Nik Taylor. 2021. 'Animals as Domestic Violence Victims: A Challenge for Humanist Social Work'. In *Post-Anthropocentric Social Work: Critical Posthuman and New Materialist Perspectives*, edited by Vivienne Bozalek and Bob Pearse, 187–97. London and New York: Routledge.

Fraser, Nancy. 2017. 'Crisis of Care? On the Social-reproductive Contradictions of Contemporary Capitalism'. In *Social Reproduction Theory: Remapping Class, Recentering Oppression*, edited by Tithi Bhattacharya. London: Pluto.

Fridlund, Alan J., and Melissa MacDonald. 1998. 'Approaches to Goldie: A Field Study of Human Responses to Canine Juvenesence'. *Anthrozoös* 11, no. 2: 95–100.

Fudge, Erica. 2002. *Animal*. London: Reaktion Books.

Fudge, Erica. 2008. *Pets*. Stocksfield: Acumen.

Gaard, Greta. 1993. 'Living Interconnections with Animals and Nature'. In *Ecofeminism: Women, Animals, Nature*, edited by Greta Gaard, 1–12, Philadelphia: Temple University Press.

Gaard, Greta. 2012. 'Speaking of Animal Bodies'. *Hypatia – Special Issue Animal Others* 27, no. 3: 520–25.

Gabb, Jacqui. 2008. *Researching Intimacy in Families*. Basingstoke: Palgrave Macmillan.

Gane, Nick, and Donna Haraway. 2006. 'When We Have Never Been Human, What Is to Be Done? An Interview with Donna Haraway'. *Theory, Culture and Society* 23, nos. 7–8: 135–58.

Garber, Marjorie. 1997. *Dog Love*. New York: Touchstone.

Gieryn, Thomas F. 2000. 'A Space for Place in Sociology.' *Annual Review of Sociology* 26: 463–96. http://www.jstor.org/stable/223453.

Gilding, Michael. 2010. 'Reflexivity Over and Above Convention: The New Orthodoxy in the Sociology of Personal Life, Formerly Sociology of the Family'. *British Journal of Sociology* 61, no. 4: 757–77.

Giraud, Eva Haifa. 2019. *What Comes After Entanglement? Activism, Anthropocentrism and an Ethics of Exclusion*. Durham, NC, and London: Duke University Press.

Gov.UK. 2023. 'Banned Dogs'. Accessed 4 July 2023. https://www.gov.uk/control-dog-public/banned-dogs#:~:text=In%20the%20UK%2C%20it's%20against,Japanese%20Tosa.

Greenbaume, Jessica. 2004. 'It's a Dogs' Life: Elevating from Pet to "Fur Baby" at Yappy Hour'. *Society and Animals* 12, no. 2: 117–35.

Gruen, Lori. 2015. *Entangled Empathy: An Alternative Ethic for Our Relationship with Animals*. New York: Lantern Books.

Guéguen, Nicholas, and Serge Ciccoti. 2008. 'Domestic Dogs as Facilitators in Social Interaction: An Evaluation of Helping and Courtship Behaviours'. *Anthrozoös* 21, no. 4: 339–49.

Guenther, Katja M. 2019. '"Taking the Ghetto Out of the Dog": Reproducing Inequality in Pit Bull Rescue'. *Ethnic and Racial Studies* 43, no. 10: 1795–812.

Guenther, Katja M. 2020. *The Lives and Deaths of Shelter Animals.* Stanford: Stanford University Press.

Gumbs, Alexis Pauline. 2016. 'M/other Ourselves: A Black Queer Feminist Genealogy for Radical Mothering'. In *Revolutionary Mothering: Love on the Front Lines*, edited by Alexis Pauline Gumbs, China Martens and Mai'a Williams, 19–31. Oakland, CA: PM Press

Gurden, Jay. 2021. *Understanding Reactive Dogs: Why Dogs React and How to Help.* UK: Amazon.

Hall, Tom, and Robin J. Smith. 2017. 'Seeing the Need: Urban Outreach as Sensory Walking'. In *Walking Through Social Research*, edited by Charlotte Bates and Alex Rhys-Taylor, 38–54. Abingdon: Routledge.

Hall, Tom, Brett Lashua and Amanda Coffey. 2006. 'Stories as Sorties'. *Qualitative Research* 3, no. 3: 2–4.

Hamilton, Lesley, and D. McCabe. 2016. '"It's Just a Job": Understanding Emotion Work, De-Animalization and the Compartmentalization of Organized Animal Slaughter'. *Organization* 23, no. 3: 330–50.

Hamilton, Lesley, and Nik Taylor. 2013. *Animals at Work: Identity, Politics and Culture in Work with Animals.* Boston and Leiden: Brill Academic Publishers.

Hamilton, Lesley, and Nik Taylor. 2017. *Ethnography After Humanism: Power, Politics and Method in Multi-Species Research.* Basingstoke: Palgrave Macmillan.

Haraway, Donna J. 1988. 'Situated Knowledges: The Science Question in Feminism and the Privilege of Partial Perspective'. *Feminist Studies* 14, no. 3: 575–99.

Haraway, Donna J. 2003. *Companion Species Manifesto.* Chicago: Prickly Paradigm Press.

Haraway, Donna J. 2008. *When Species Meet.* London and Minneapolis: University of Minnesota Press.

Haraway, Donna J. 2015. 'Anthropocene, Capitalocene, Plantationocene, Chthulucene: Making Kin'. *Environmental Humanities Review* 6: 159–65.

Haraway, Donna J. 2016a. *Staying with the Trouble: Making Kin in the Chthulucene.* Durham and London: Duke University Press.

Haraway, Donna J. 2016b. *Manifestly Haraway.* Minneapolis: University of Minnesota Press.

Haraway, Donna J. 2017. 'Making Oddkin: Telling Stories for Earthly Survival'. Talk at Duke Franklin Humanities Institute, Duke University, 31 October. Accessed 24 November 2022. https://www.youtube.com/watch?v=rMBRX9EcrH8.

Haraway, Donna J. 2018. 'Making Kin in the Chthulucene: Reproducing Multispecies Justice'. In *Making Kin Not Population*, edited by Adele E. Clarke and Donna J. Haraway, 67–100. Chicago, IL: Prickly Paradigm Press.

Haraway, Donna J. 2022. 'Speculative Fabulations for Technoculture's Generations: Taking Care of Unexpected Country'. In *Kin: Thinking with Deborah*

Bird Rose, edited by Tom van Dooren and Matthew Chrulew, 70–93. Durham, NC: Duke University Press. https://doi.org/10.2307/j.ctv29z1h77.7.

Harringtons. 2021. 'Harringtons: The Natural Choice' TV Advert. YouTube video, posted 17 January. https://www.youtube.com/watch?v=KlDyYlhvrF4.

Harris. 2015. 'More than Ever, Pets Are Members of the Family'. Harris Poll. Accessed 9 November 2022. https://theharrispoll.com/whether-furry-feathered-or-flippers-a-flapping-americans-continue-to-display-close-relationships-with-their-pets-2015-is-expected-to-continue-the-pet-industrys-more-than-two-decades-strong/.

Harvey, David. 1996. *Justice, Nature and the Geography of Difference*. Oxford: Blackwell.

Hearne, Vicki. 2000. *Adams Task: Calling Animals by Name.* Los Angeles: Adakin Press.

Hearne, Vicki. 2007. *Bandit: The Heart-warming Story of One Dogs Rescue from Death Row*. New York: Skyhorse Publishing.

Hediger, Ryan, ed. 2012. *Animals and War: Studies of Europe and North America.* Lieven: Brill.

Henderson, A., Sandra Harmon and Harmony Newman. 2016. 'The Price Mothers Pay, Even When They Are Not Buying It: Mental Health Consequences of Idealized Motherhood'. *Sex Roles* 74, nos. 11–12: 512–26.

Hester, Helen. 2018. *Care Under Capitalism: The Crisis of 'Women's Work'. IPPR Progressive Review* 24, no. 4: 343–52.

Hetherington, Kevin. 1997. *The Badlands of Modernity: Heterotopia and Social Ordering*. Abingdon: Routledge.

Hochschild, Arlie. 1997. *The Ties That Bind: When Work Becomes Home and Home Becomes Work*. New York: Metropolitan Books.

Hochschild, Arlie. 2012 [1983]. *The Managed Heart: Commercialisation of Human Feeling*. Berkely: University of California Press.

Hohti, Riikka. and Tuure Tammi. 2019. 'The Greenhouse Effect: Multispecies Childhood and Non-innocent Relations of Care'. *Childhood* 26, no. 2: 169–85. https://doi.org/10.1177/0907568219826263.

hooks, bell. 1984. *Feminist Theory: From Margin to Center*. Boston: South End Press.

hooks, bell. 1990. *Yearning: Race, Gender and Cultural Politics,* Boston, MA: Beacon Press.

Hribal, Jason. 2003. '"Animals Are Part of the Working Class": A Challenge to Labor History'. *Labour History* 44, no. 4: 435–53.

Hurn, Samantha. 2010. 'What's in a Name? Anthrozoology, Human-Animal Studies, Animal Studies or . . .?' *Anthropology Today* 26, no. 3: 27–28.

Hurn, Samantha. 2012. *Humans and Other Animals: Cross-Cultural Perspectives on Humananimal Interactions*. London: Pluto Press.

Iams. 2015. Iams Cat Food Commercial 'These Teeth Were Made for Meat'. Accessed 31 March 2023. https://www.youtube.com/watch?v=o9_wxYfWXqw.

Ingold, Tim. 2012. *The Perception of the Environment: Essays on Livelihood, Dwelling and Skill*. London: Routledge.

Irvine, Leslie. 2004. *If You Tame Me: Understanding Our Connection with Animals*. Philadelphia: Temple University Press.

Irvine, Leslie. 2008. 'Animals and Sociology'. *Sociology Compass* 2, no. 6: 1954–71.

Irvine, Leslie. 2013a. 'Animals as Lifechangers and Lifesavers: Pets in the Redemption Narratives of Homeless People'. *Journal of Contemporary Ethnography* 42, no. 1: 3–30.

Irvine Leslie. 2013b. *My Dog Always Eats First: Homeless People and Their Animals*. Boulder CO: Lynne Rienner Publishers.

Irvine, Leslie, and Lauren Cilia. 2016. *The Value of Pets to Public and Private Health and Well-Being*. https://animalstudiesrepository.org/humcarel.

Irvine, Leslie, and Lauren Cilia. 2017. 'More-than-Human Families: Pets, People and Practices in Multispecies Households'. *Sociology Compass* 11, no. 2: e12455.

Jamieson, Lynn. 1998. *Intimacy: Personal Relationships in Modern Society*. Cambridge: Polity.

Jerolmack, Colin. 2005. 'Our Animals, Our Selves? Chipping Away the Human-Animal Divide'. *Sociological Forum* 20, no. 4: 651–60.

Jones, Phil, Griff Bunce, James Evans, Hannah Gibbs and Jane Ricketts-Hein. 2008. 'Exploring Space and Place with Walking Interviews'. *Journal of Research Practice*, 4, no. 2: 150–61. https://files.eric.ed.gov/fulltext/EJ827010.pdf.

Joy, Melanie. 2010. *Why We Love Dogs, Eat Pigs and Wear Cows: An Introduction to Carnism*. Newburyport, MA: Conari Press.

Kateryna, Miliutina, Andrey Trofimov, Zelenin Vsevolod, Tetiana Andrushchenko and Liudmyla Karamushka. 2022. 'The Role of Pets in Preserving the Emotional and Spiritual Wellbeing of Ukrainian Residents During Russian Hostilities'. *Journal of Religion and Health* 62: 500–509. https://doi.org/10.1007/s10943-022-01669-4.

Kelly, Kieran. 2022. 'Fury Over Illegal Parties Where Ravers "Blast Drum and Bass for 9 Hours a Night" and Leave Behind "Human Poo"'. *My London*, 17 May. Accessed 9 March 2023. https://www.mylondon.news/news/east-london-news/mum-furious-illegal-parties-blasting-23968309.

Kennel Club UK. 2013. 'Dog Walking'. Accessed 25 June 2013. http://www.thekennelclub.org.uk/item/2712.

Kim, Claire Jean. 2015. *Dangerous Crossings: Races, Species, and Nature in a Multicultural Age*. Cambridge: Cambridge University Press.

Kirksy, S. Eben, and Stefan Helmreich. 2010. 'The Emergence of Multispecies Ethnography'. *Cultural Anthropology* 25, no. 4: 545–76.

Knight, Andrew. 2005. 'In Defense of Vegetarian Cat Food', *Journal of the American Veterinary Association* 226, no. 4: 512–13.

Knight, Andrew. 2008. 'Fishy Business?' *Lifescape* (May): 74–76.

Knight, Andrew, Eason Huang, Nicholas Rai and Hazel Brown. 2022. 'Vegan Versus Meat-based Dog Food: Guardian-reported Indicators of Health'. *PLoS ONE* 17, no. 4: e0265662. https://doi.org/10.1371/journal.pone.0265662.

Knight, Andrew, and Leitsberger, Madelaine. 2016. 'Vegetarian Versus Meat-Based Diets for Companion Animals' *Animals* 6, no. 9: 57. https://doi.org/10.3390/ani6090057.

Knight, Sarah, and Victoria Edwards. 2008. 'In the Company of Wolves: The Physical, Social and Psychological Benefits of Dog Ownership'. *Journal of Aging and Health* 20: 437–55.

Koontz, Dean. 1988. *Oddkins: A Fable for All Ages*. New York: Warner Books.

Kopnina, Helen. 2017. 'Beyond Multispecies Ethnography: Engaging with Violence and Animal Rights in Anthropology'. *Critique of Anthropology* 37, no. 3: 333–57.

Korać, Maja. 2009. *Remaking Home: Reconstructing life, place and identity in Rome and Amsterdam*. New York and Oxford: Berghahn Books.

Kusenbach, Margarethe. 2003. 'Street Phenomenology: The Go-Along as an Ethnographic Research Tool'. *Ethnography* 4, no. 3: 455–85.

Kusenbach, Margarethe, and Krista E. Paulsa, eds. 2013. *Home: International Perspectives on Culture, Identity and Belonging*. New York: Peter Laing.

Laing, Melissa. 2021a. 'On Being Posthuman in Human Spaces: Critical Posthumanist Social Work with Interspecies Families'. *International Journal of Sociology and Social Policy* 41, no. 3/4: 361–65.

Laing, Melissa. 2021b. 'Encountering Interspecies Homelessness: Resisting Anthroparchy in Social Work and the All-Too-human Services'. In *Post-Anthropocentric Social Work: Critical Posthuman and New Materialist Perspectives*, edited by Vivienne Bozalek and Bob Pearse, 187–97. London and New York: Routledge.

Laing, Melissa, and Christopher Mayela. 2018. '"They Burn Brightly, But Only for a Short Time": The Role of Social Workers in Companion Animal Grief and Loss'. *Anthrozoös* 31, no. 2: 221–32. https://doi.org/10.1080/08927936.2018.1434062.

Lancendorfer, Karen M., Jo-Ann L. Altkin and Bonnie B. Reece. 2008. 'Animals in Advertising: Love Dogs? Love the Ad!'. *Journal of Business Research* 6, no. 5: 384–91.

Lane, Karen. 2015. 'Canine Connections: Fieldwork with a Dog as Research Assistant'. *Anthropology in Action* 22, no. 3: 27–38.

Lau, Edie. 2011. 'Pet Owners Receive $12.4 Million in Melamine Case'. *Vin News*, 12 October. http://news.vin.com/VINNews.aspx?articleId=20025.

Laurent-Simpson, Andrea. 2017. '"They Make Me Not Wanna Have a Child": Effects of Companion Animals on Fertility Intentions of the Childfree'. *Sociological Inquiry* 87, no. 4: 586–607.

Laurent-Simpson, Andrea. 2021. *Just Like Family: How Companion Animals Joined the Household*. New York: New York University Press.

Lasch, Christopher. 1995 [1977]. *Haven in a Heartless World: The Family Besieged*. London: W.W. Norton and Co.

Lavin, Chad. 2013. *Eating Anxiety: The Perils of Food Politics*. Minneapolis: University of Minnesota Press.

Lescureux, Nicolas, and John D. C. Linnell. 2014. 'Warring Brothers: The Complex Interactions Between Wolves (Canis Lupus) and Dogs (Canis Familiaris) in a Conservation Context'. *Biological Conservation* 171 (March): 232–45. https://doi.org/10.1016/j.biocon.2014.01.032.

Levi Martin, John. 2000. 'What Do Animals Do All Day?: The Division of Labour, Class Bodies and Totemic Thinking in the Popular Imagination'. *Poetics* 27, no. 2/3: 195–231.

Lukes, Steven. 1973. *Emile Durkheim: His Life and Work – An Historical and Critical Study*. New York: Harper & Row.

MacCormack, Patricia. 2020. *The Ahuman Manifesto: Activism for the End of the Anthropocene*. London: Bloomsbury Academic.

Madigan, Ruth, and Moira Munro. 1996. 'House Beautiful: Style and Consumption in the Home'. *Sociology* 30, no. 1: 41–57.

Madigan, Ruth, Moira Munro and Susan J. Smith. 1990. 'Gender and the Meaning of Home'. *International Journal of Urban and Regional Research* 14, no. 4: 625–47.

Maher, Jennifer, and Harriet Pierpoint. 2011. 'Friends, Status Symbols and Weapons: the Use of Dogs by Youth Groups and Youth Gangs'. *Crime, Law and Social Change* 55, no. 5: 405–20. https://doi.org/10.1007/s10611-011-9294-5.

Mallett, Shelley. 2004. 'Understanding Home: A Critical Review of the Literature'. *The Sociological Review* 52, no. 1: 62–89. https://doi.org/10.1111/j.1467-954X.2004.00442.x.

Marvin, Garry. 2005. 'Guest Editors Introduction – Seeing, Looking, Watching, Observing Nonhuman Animals'. *Society and Animals* 13, no. 1: 1–12.

Marvin, Garry. 2006. 'Wild Killing: Contesting the Animal in Hunting'. In *Killing Animals*, edited by the Animal Studies Group, 10–29. Urbana and Chicago: University of Illinois Press.

Mason, Jennifer, and Becky Tipper. 2008. 'Being Related: How Children Define and Create Kinship'. *Childhood* 15, no. 4: 441–60.

Massey, Doreen. 1994. *Space, Place and Gender*. Cambridge: Polity.

Massey, Doreen. 1995. *For Space*. Cambridge: Polity.

Masson, Jeffrey Moussaieff. 1998. *Dogs Never Lie about Love*. London: Vintage.

Mattheis, Nikolas. 2022. 'Making Kin, Not Babies? Towards Childist Kinship in the "Anthropocene"'. *Childhood* 29, no. 4: 512–28. https://doi.org/10.1177/09075682221098155.

McCabe, Darren, and Lindsay Hamilton. 2015. 'The Kill Programme: An Ethnographic Study of the "Dirty Work" of Meat Inspectors in a Slaughterhouse'. *New Technology, Work and Employment* 30, no. 2: 95–108.

McColgan, Gillian, and Irene Schofield. 2007. 'The Importance of Companion Animal Relationships in the Lives of Older People'. *Nursing Older People* 19, no. 1: 21–23. doi:10.7748/nop2007.02.19.1.21.c4361.

McConnell, Allen R., Christina M. Brown, Tonya M. Shoda, Laura E. Stayton and Colleen E. Martin. 2011. 'Friends with Benefits: On the Positive Consequences of Pet Ownership'. *Journal of Personal Social Psychology* 101, no. 6: 1239–52. https://psycnet.apa.org/doi/10.1037/a0024506.

McConnell, Patricia B. 2002. *The Other End of the Leash: Why We Do What We Do Around Dogs.* New York: Ballantine Books.

McConnell, Patricia B. 2007. *For the Love of a Dog: Understanding Emotion in You and Your Best Friend.* New York: Ballantine Books.

McHugh, Susan. 2004. *Dog*. London: Reaktion Books.

McNicholas, June, and Glyn M. Collis. 2000. 'Dogs as Catalysts for Social Interactions: Robustness and Effect'. *British Journal of Psychology* 91, no. 1: 61–70. https://doi.org/10.1348/000712600161673.

Mellor, Mary. 1992. *Breaking the Boundaries: Towards a Feminist Green Socialism*. London: Virago.

Mellor, Mary. 1997. *Feminism and Ecology*. Cambridge: Polity.

Mercer, David. 2021. 'Hundreds of "Dangerous" Dogs Killed After Being Seized by Police – but Fears Many "Needlessly Euthanised"'. *Sky News*, 26 September. https://news.sky.com/story/hundreds-of-dangerous-dogs-killed-after-being-seized-by-police-but-fears-many-needlessly-euthanised-12498070#:~:text=At%20least%205%2C333%20dogs%20have,Freedom%20of%20Information%20Act%20show.

Messent, Peter, R. 1983. 'Social Facilitation of Contact with Other People by Pet Dogs'. In *New Perspectives in Our Lives with Companion Animals*, edited by Aaron H. Katcher and Alan M. Beck, 37–46. Philadelphia: University of Philadelphia Press.

Miah, Andy. 2007. 'Posthumanism: A Critical History'. In *Medical Enhancement and Posthumanity*, edited by Bert Gordijn and Ruth Chadwick, 71–94, London: Routledge.

Middleton, Joe. 2022. '"Out of Control Rottweiler" Attacked 5 People in Park after It Was off Lead'. *The Independent*, 5 September 2022. https://www.independent.co.uk/news/uk/home-news/luton-park-dog-attack-rottweiler-b2159408.html.

Molloy, Claire. 2011. 'Dangerous Dogs and the Construction of Discourses of Risk'. In *Theorizing Animals*, edited by Nik Taylor, 107–28. Brill: Leiden.

Morgan, David. 1996. *Family Connections: An Introduction to Family Studies*. Cambridge: Polity Press.

Morgan, Karen, and Matthew Cole. 2011. 'The Discursive Representation of Non-human Animals in a Culture of Denial'. In *Human and Other Animals: Critical Perspectives*, edited by Bob Cater and Nickie Charles, 112–32. London: Palgrave Macmillan.

Motamedi-Fraser, M. 2019. 'Dog Words – Or, How to Think Without Language'. *The Sociological Review* 67, no. 2: 374–90.

Motamedi-Fraser, Mariam. 2024. *Dog Politics: Species Stories in the Animal Sciences*. Manchester: Manchester University Press.

Mouret, Sebastian. 2017. 'Apprendre à prendre soin: La centralité du travail dansl'éducation des chiens guides d'aveugles'. *Ecologie et politique* 54, no. 1: 87–102. https://www.cairn.info/revue-ecologie-et-politique-2017-1-page-87.htm.

Murphy, Michelle. 2017. *The Economization of Life*. Durham, NC: Duke University Press.

Murray, Jessica. 'American XL Bully Dog Will Be Banned, Says Rishi Sunak', *The Guardian*, 15 September 2023, https://www.theguardian.com/lifeandstyle/2023/sep/15/american-xl-bully-dog-will-be-banned-says-sunak.

Murray, Jessica. 'Britons Have Started Handing In XL Bullies to Be Put Down, MPs Told', *The Guardian*, 18 October 2023, https://www.theguardian.com/uk-news/2023/oct/18/britons-have-started-handing-in-xl-bullies-to-be-put-down-mps-told.

Naess, Arne. 1979. 'Self-Realization in Mixed Communities of Humans, Bears, Sheep and Wolves'. *Inquiry* 22, nos.1–4: 231–41.

Neil, Sarah, and Karim Murji. 2015. 'Sociologies of Everyday Life: Editors Introduction, Special Issue Sociology of Everyday Life'. *Sociology* 49, no. 5: 811–19.

Nestle, Marion. 2008. *Pet Food Politics: The Chihuahua in the Coal Mine*. London: University of California Press.

Newman, Oscar. 1972. *Defensible Space: Crime Prevention through Urban Design*. London: Macmillan.

Nibert, David. 2002. *Animal Rights/Human Rights: Entanglements of Oppression and Liberation*. Lanham, MD: Rowman & Littlefield.

Nibert, David. 2013. *Animal Oppression and Human Violence: Domesecration, Capitalism and Global Conflict*. New York: Colombia University Press.

Nibert, David, ed. 2017. *Animal Oppression and Capitalism*, vol. 1, *The Oppression of Nonhuman Animals as Sources of Food*. Denver, CO: Praeger Press.

Nocella II, Anthony J., John Sørenson, Kim Socha and Atsuko Matsuoka, eds. 2014. *Defining Critical Animal Studies: An Intersectional and Social Justice Approach for Liberation*. Peter Lang: New York.

Nocella II, Anthony J., and Kim Socha, eds. 2022 *Radical Animal Studies: Beyond Respectability Politics, Opportunism and Cooptation*. New York: Peter Lang.

Noske, Barbara. 1997. *Beyond Boundaries: Humans and Animals*. Montreal: Black Rose Books.

Nugent, William R., and Linda Daugherty. 2022. 'A Measurement Equivalence Study of the Family Bondedness Scale: Measurement Equivalence Between Cat and Dog Owners'. *Frontiers of Veterinary Science* 8: 812–922. https://www.frontiersin.org/articles/10.3389/fvets.2021.812922/full.

Oakley, Anne. 1974. *The Sociology of Housework*. London: John Wiley & Sons.

Offe, Claus. 1985. *Disorganised Capitalism: Contemporary Transformations of Work and Politics*. Cambridge: Polity.

Ojeda, Diane, Jade S. Sasser and Elizabeth Lunstrum. 2020. 'Malthus's Specter and the Anthropocene'. *Gender, Place & Culture* 27, no. 3: 316–32.

Oldenburg, Ray. 1999. *The Great Good Place: Cafés, Coffeeshops, Bookstores, Bars, Hair Salons and Other Hangouts at the Heart of a Community*. New York: Marlow and Company.

Oliver, Catherine. 2021. 'Beyond-Human Ethics: The Animal Question in Institutional Ethical Reviews'. *Area* 53, no. 4: 619–26.

Owens, Nicole, and Liz Grauerholz. 2019. 'Interspecies Parenting: How Pet Parents Construct Their Roles'. *Human Society* 43, no. 2: 96–119.

Palmer, Claire. 2006. 'Killing Animals in Animal Shelters'. In *Killing Animals*, edited by the Animal Studies Group, 170–87. Urbana and Chicago: University of Illinois Press.

PAT (Pets as Therapy). 2020. *Factsheet 3 PAT Temperament Assessment*. https://petsastherapy.org/wp-content/uploads/2018/08/Factsheet3-PATTemperament-Assessment-Test-dog.pdf.

Paulson, Steve. 2019. 'Making Kin: An Interview with Donna Haraway'. *New York Review of Books*, 6 December 2019. https://lareviewofbooks.org/article/making-kin-an-interview-with-donna-haraway/.

Peden, James. 1999. *Vegetarian Cats and Dogs*, third edition. Troy, MT: Harbingers of a New Age.

Peggs, Kay. 2012. *Animals and Sociology*. Basingstoke: Palgrave.

Peggs, Kay. 2014. 'From Centre to Margins (and Back Again): Critical Animal Studies and the Reflexive Human Self'. In *From the Margins to the Centre? The Rise of Critical Animal Studies*, edited by Nik Taylor and Richard Twine, 36–51. Routledge: London.

PFMA (Pet Food Manufacturers Association, now UK Pet Food). 2014. 'Pet Population 2013'. http://www.pfma.org.uk/petpopulation.

PFMA. 2022. 'Pet Population 2021'. https://www.pfma.org.uk/pet-population-2021.

Pick, Anat. 2011. *Creaturely Poetics: Animality and Vulnerability in Literature and Film.* New York: Colombia University Press.

Price, Margaret. 2017. 'What Is a Service Animal? A Careful Rethinking'. *Review of Disability Studies* 13, no. 4: 1–18. https://rdsjournal.org/index.php/journal/article/view/757/1905.

Pet Health Council. 2019. 'People and Pets'. http://www.pethealthcouncil.co.uk/people-and-pets.

Philo, Chris, and Chris Wilbert, eds. 2000. *Animal Spaces, Beastly Places: New Geographies of Human-Animal Relations*. London and New York: Routledge.

Pierce, Jessica. 2012. *The Last Walk: Reflections on Our Pets at the End of Their Lives*. Chicago and London: Chicago University Press.

Pierce, Jessica. 2016. *Run, Spot, Run: The Ethics of Keeping Pets.* Chicago: Chicago University Press.

Pierce, Jessica, and Marc Bekoff. 2021. *A Dog's World: Imagining the Lives of Dogs in a World without Humans*. Princeton and Oxford: Princeton University Press.

Pink, Sarah. 2012. *Situating Everyday Life.* London: Sage.

Plumwood, Val. 2002. *Environmental Culture: The Ecological Crisis of Reason.* London: Routledge.

Porcher, Jocelyn. 2011. *Vivre avec les Animaux: Une Utopie pour le XXI Ème*. Paris: La Découverte.

Porcher, Joceline. 2014. 'The Work of Animals: A Challenge for the Social Sciences'. *Humanimalia* 6, no. 1: 1–9.

Porcher, Jocelyn, and Elisabeth Lécrivain. 2019. 'The Wolf and the Patou Dog: Freedom and Work'. In *Animal Work: A New Perspective on Human-Animal Relations*, edited by Jocelyn Porcher and Jean Estebanes. Bielefield: Transcript Verlag. https://doi.org/10.1515/9783839443644.

Porcher, Joceline, and Tiphainie Schmitt. 2012. 'Dairy Cows: Workers in the Shadows?' *Society and Animals* 20, no. 1: 39–60.

Potts, Annie, and Donna Haraway. 2010. 'Kiwi Chicken Advocate Talks with Californian Dog Companion'. *Feminism and Psychology* 20, no. 3: 318–36.

Powell, Lauren, Deborah L. Duffy, Katherine A. Kruger, Brittany Watson and James A. Serpell. 2021. 'Relinquishing Owners Underestimate Their Dogs Behavioural

Problems: Deception or Lack of Knowledge?' *Frontiers in Veterinary Medical Science* 8. https://www.frontiersin.org/articles/10.3389/fvets.2021.734973/full.

Power, Emma. 2008. 'Furry Families: Making a Human-dog Family through Home'. *Social and Cultural Geography* 9, no. 5: 535–55.

Power, Emma R. 2017. 'Renting with Pets: A Pathway to Housing Insecurity?'. *Housing Studies* 32, no. 3: 336–60. https://doi.org/10.1080/02673037.

Procter & Gamble press release, 2012. 'New Campaign Inspired by Real Stories from Pet Owners'. Accessed on 8 July 2014. http://news.pg.com/press-release/pg-corporate-announcements/iams-launches-new-campaign-inspired-real-stories-showcase-e#sthash.FH2P62g0.dpuf.

Puig de la Bellacasa, Maria. 2012. '"Nothing Comes Without Its World": Thinking with Care'. *The Sociological Review* 60, no. 2: 197–216.

Puig de la Bellacasa, Maria. 2017. *Matters of Care: Speculative Ethics in More Than Human Worlds*. Minneapolis: University of Minnesota Press.

Pure Pet Food. 2022. Television advert. Accessed 31 March 2023. https://www.youtube.com/watch?v=XqwdnxUXoW8.

Pure. 2020. 'Pure Pet Food 2022 TV Advert'. YouTube video, posted 8. Accessed 30 March 2023. https://www.youtube.com/watch?v=lfls5c-Z7bs.

Redmalm, David. 2014. 'Holy Bonsai Wolves: Chihuahuas and the Paris Hilton Syndrome'. *International Journal of Cultural Studies* 17, no. 1: 93–109.

Redmalm, David. 2015. 'Pet Grief: When Is Non-Human Life Grievable?' *The Sociological Review* 63, no. 1: 19–35.

Redmalm, David. 2021. 'Discipline and Puppies: The Powers of Pet Keeping'. *International Journal of Sociology and Social Policy* 41, no. 3/4: 440–54.

Reeves, Matthew J., Anne P. Rafferty, Corinne E. Miller and Sarah K. Lyon-Callo. 2011. 'The Impact of Dog Walking on Leisure Time Physical Activity: Results from a Population Survey of Michigan Adults'. *Journal of Physical Activity and Health* 8, no. 3: 436–44.

Relph, Edward. 1976. *Place and Placelessness*. London: Pion.

Ritzer, George. 1992. *Classical Sociological Theory*. New York: McGraw Hill.

Rose, Gillian. 1993. *Feminism and Geography: The Limits of Geographical Knowledge*. Cambridge: Polity.

Rosenberg, Meisha. 2011. 'Golden Retrievers Are White, Pit Bulls Are Black and Chihuahuas Are Hispanic: Representations of Breeds of Dog and Issues of Race in Popular Culture'. In *Making Animal Meaning*, edited by Linda Kalof and Georgina M. Montgomery, 113–26. East Lansing: Michigan State University Press.

Rosiek, Jerry Lee, Jimmy Snyder and Scott L. Pratt. 2020. 'The New Materialisms and Indigenous Theories of Non-Human Agency: Making the Case for Respectful Anti-Colonial Engagement'. *Qualitative Inquiry* 26, no. 3/4: 331–46.

Rosling, Elisabeth. 2022 'A Guide to New Dog Laws and Rules for Owners, UK 2022'. *Barc London*, June. https://www.barclondon.com/blogs/guides/new-dog-laws-uk-2022#:~:text=The%20code%20highlights%20the%20importance,the%20time%20of%20your%20walk.

Rowlands, Mark. 2009. 'Tom Regan: Animal Rights as Natural Rights'. In *Animal Rights: Moral Theory and Practice*, 58–97. Basingstoke: Palgrave.

RSPCA. 2023. 'Advice and Welfare/Fish'. https://www.rspca.org.uk/adviceandwelfare/pets/fish.

RSPCA. 2019. 'Our Bid to Help 7m Dogs Suffering from Separation Anxiety'. https://www.rspca.org.uk/whatwedo/latest/details/-/articleName/2019_03_06_7m_dogs_could_be_suffering_from_separation_anxiety.

RSPCA. 2014. 'Breed Specific Legislation: Don't Believe the Type!'. http://www.rspca.org.uk/getinvolved/campaigns/dogownership/bsl.

RSPCA. n.d. 'Handle with Care: Stick Insects'. https://www.rspca.org.ukstick+insects+(PDF+124KB).pdf.

Saldhana, Arun. 2008. 'Heterotopia and Structuralism'. *Environment and Planning A: Economy and Space* 40, no. 9: 2080–96.

Sanders, Clinton. 1993. 'Understanding Dogs: Caretakers' Attributions of Mindedness in Canine-Human Relationships'. *Journal of Contemporary Ethnography* 22, no. 2: 205–26.

Sanders, Clinton. 1999. *Understanding Dogs: Living and Working with Canine Companions.* Philadelphia: Temple University Press.

Sanders, Clinton, and Arnold Arluke. 1993. 'If Lions Could Speak: Investigating the Animal-human Relationship and the Perspectives of Nonhuman Others'. *Sociological Quarterly* 34, no. 3: 377–90.

Sanders, Peter. 1990. *A Nation of Homeowners*. London: Unwin Hyman.

Satama, Suvi, and Astrid Huopalainen. 2019. '"Please Tell Me When You Are in Pain": A Heartbreaking Story of Care, Grief and Female-Canine Companionship'. *Gender, Work and Organization* 26, no. 3: 358–76.

Savage, Mike, Gaynor Bagnall and Brian Longhurst. 2005. *Globalization and Belonging: The Suburbanization of Identity*. London: Sage.

Sayers, Janet, Rachel Forrest and Maria Pearson. 2022. 'Furry Families: Ethical Entanglements through More-than-Human Domestic Dramas'. *Sociological Research Online* 28, no. 2: 540–57. https://doi.org/10.1177/13607804211064913.

Scanlon, Louise, Pru Hobson-West, Kate Cobb, Ann McBride and Jenny Stavisky. 2021. 'Homeless People and Their Dogs: Exploring the Nature and Impact of the Human–Companion Animal Bond'. *Anthrozoös* 34, no. 1: 77–92.

Schleis, Paul. 'The Drug that Became Prescription for Trouble', The Washington Post, 8 May 2005, https://www.washingtonpost.com/archive/lifestyle/2005/05/08/the-drug-that-became-prescription-for-trouble/2ebf91cc-a16c-4b38-881c-0c4112e8aa0c/.

Herald Scotland. 'Terror as Rottweilers Attack Pupils', 14 February 1990, https://www.heraldscotland.com/news/11974280.terror-as-rottweilers-attack-pupils/ and 'School Attack Dogs to Be Destroyed', 10 May 1990, https://www.heraldscotland.com/news/11944251.school-attack-dogs-to-be-destroyed/.

Segerdahl, Pär. 2012. 'Humanizing Nonhumans: Ape Language Research as Critique of Metaphysics'. In *Language, Ethics and Animal Life: Wittgenstein and Beyond*, edited by Niklas Forsberg, Mikel Burley and Nora Hämäläinen, 16–31. New York and London: Bloomsbury Academic.

Serpell, James. 1991. 'Beneficial Effects of Pet Ownership on Some Aspects of Human Health and Behaviour'. *Journal of the Royal Society of Medicine* 84, no. 12: 717–20.

Serpell, James. 1996. *In the Company of Animals: A Study of Human-Animal Relat*ionships. Cambridge: Cambridge University Press.

Serpell, James, and E. Paul. 1994. 'Pets and the Development of Positive Attitudes Towards Animals'. In *Animals and Society: Changing Perspectives*, edited by Aubrey Manning and James Serpell, 127–44. London: Routledge.

Shapiro, Kenneth. 1990. 'Understanding Dogs Through Kinasthetic Empathy, Social Construction and History'. *Anthrozoös* 3, no. 3: 184–95.

Shapiro, Kenneth. 2002. 'Editors Introduction – The State of Human-Animal Studies: Solid at the Margin!'. *Society and Animals* 10, no. 4: 331–37.

Shapiro, Kenneth. 2020. 'Human-Animal Studies: Remembering the Past, Celebrating the Present, Troubling the Future'. *Society and Animals* 28, no. 7: 797–833.

Shaw, Deborah Benita. 2018. *Posthuman Urbanism: Mapping Bodies in Contemporary City Space*. Lanham, MD: Rowman & Littlefield.

Shoard, Marion. 2002. 'Edgelands'. In *Remaking the landscape: The Changing Face of Britain*, edited by Jennifer Jenkins, 117–46. London: Profile Books.

Shukin, Nicole. 2013. 'Security Bonds: On Feeling Power and the Fiction of an Animal Governmentality'. *ESC: English Studies in Canada* 39, no. 1: 177–98.

Sibley, David. 1995a. *Geographies of Exclusion: Society and difference in the West*. London: Routledge.

Sibley, David. 1995b. 'Families and Domestic Routines: Constructing the Boundaries of Childhood'. In *Mapping the Subject: Cultural Geographies of Transformation*, edited by Steve Pile and Nigel Thrift, 123–40. London: Routledge.

Simmel, Georg. 1972. *Georg Simmel: On Individuality and Social Forms.* Edited by Donald N. Levine. Chicago: University of Chicago.

Simpson, Leanne B. 2017. *As We Have Always Done: Indigenous Freedom Through Radical Resistance*. Minneapolis: University of Minnesota Press.

Singer, Peter. 1990. *Animal Liberation*, second edition. New York: Avon Books.

Sipra, Hassaan. 2022. 'Maneli Pets of South Africa Focuses on Novel Proteins Native to South Africa, Now Working with Local Customers as Well as Internationally'. *Petfood Industry*. https://www.petfoodindustry.com/articles/11833-maneli-pets-of-south-africa-recovers-to-pre-covid-revenues.

Skogland, A., and David Redmalm. 2017 'Doggy-Biopolitics: Governing via the First Dog'. *Organization* 24, no. 2: 240–66.

Smart, Carol. 2007. *Personal Life: New Directions in Sociological Thinking*. Cambridge: Polity.

Smart, Carol, and Bren Neale. 1999. *Family Fragments?* Cambridge: Polity.

Smith, Helen, Mara Miele, Nickie Charles and Rebekah Fox. 2021. 'Becoming with a Police Dog: Training Technologies for Bonding'. *Transactions of the Institute of British Geographers* 6, no. 2: 478–94.

Smith, Julie Ann. 2003. 'Beyond Dominance and Affection: Living with Rabbits in Post-Humanist Households'. *Society and Animals* 11, no. 2: 182–97.

Smith, Lauren M., Conor Goold, Rupert J. Quinnell, Alexandru M. Munteanu, Sabine Hartmann, Paolo Dalla Villa and Lisa M. Collins. 2022. 'Population Dynamics of Free-Roaming Dogs in Two European Regions and Implications for Population Control'. *PLoS ONE* 17, no. 9: e0266636. https://doi.org/10.1371/journal.pone.0266636.

Socha, Kim. 2012. *Women, Destruction and the Avant-Garde: A Paradigm for Animal Liberation*. Amsterdam: Rodolpi.

Soja, Edward W. 1996. *Thirdspace: Journeys to Los Angeles and Other Real-and-imagined Places*. Oxford: Blackwell Publishing

Sommerville, Peter. 1992. 'Homelessness and the Meaning of Home: Rooflessness or Rootlessness'. *International Journal of Urban and Regional Research* 16, no. 4: 529–39.

Specia, Mega, and Gaia Pianigiani. 2017. 'Paid Leave When Your Dog Is Sick? Sì, Italian Employer Says'. *New York Times*, 12 October. Accessed 4 November 2022. https://www.nytimes.com/2017/10/12/world/europe/italy-paid-dog-care-leave.html.

Stănescu, Vasile. 2013. 'Why "Loving" Animals is Not Enough: A Response to Kathy Rudy, Locavorism, and the Marketing of "Humane" Meat'. *The Journal of American Culture* 36, no. 2: 100–10.

Statista. 2021. 'Reasons for Dog Ownership in the UK'. https://www.statista.com/statistics/797008/reasons-for-dog-ownership-united-kingdom-uk/.

Statista. 2022a. 'Share of Households Owning a Pet in the United Kingdom (UK) from 2010/11 to 2021/22'. Accessed 15 February 2023. https://www.statista.com/statistics/308235/estimated-pet-ownership-in-the-united-kingdom-uk/.

Statista. 2022b. 'Number of Pet Dogs in the United Kingdom (UK) from 2010/11 to 2021/22' (in millions). Accessed 5 December 2022. https://www.statista.com/statistics/515379/dogs-population-in-the-united-kingdom-uk/.

Statista. 2022c. 'Share of Households in the European Union (EU) Owning at Least One Dog, by Country'. Accessed 5 December 2022. https://www.statista.com/statistics/515475/dog-ownership-european-union-eu-by-country/.

Statista. 2022d. 'Worldwide Pet Ownership by Country'. Accessed 5 December 2022. https://www.statista.com/statistics/961098/worldwide-pet-ownership-by-type-by-country/.

Statista. 2022e. 'Leading Pets Ranked by Household Ownership in the United Kingdom'. Accessed 29 May 2023. https://www.statista.com/statistics/308218/leading-ten-pets-ranked-by-household-ownership-in-the-united-kingdom-uk/.

Sundman, Ann-Sophie, Enya Van Poucke, Ann-Charlotte Svensson Holm, Åshild Faresjö, Elva Theodorsson, Per Jensen and Linda S. V. Roth. 2019. 'Long-Term Stress Levels Are Synchronized in Dogs and Their Owners'. *Scientific Reports* 9: 7391. https://doi.org/10.1038/s41598-019-43851-x.

Swabe, Joanna. 2009. 'Veterinary Dilemmas'. In *Between the Species: Readings in Human-Animal Relations*, edited by Arnold Arluke and Clinton Saunders, 248–57. New York: Pearson.

Tails.com. 2021. 'Tails.com New TV Ad'. YouTube video, posted 4 January 2019. https://www.youtube.com/watch?v=k82LAHHj6c0.

TallBear, Kim. 2018. 'Making Love and Relations: Beyond Settler Sex and Family'. In *Making Kin not Population*, edited by Adele E. Clarke and Donna J. Haraway, 145–66, Chicago, IL: Prickly Paradigm Press.

Taylor, Affrica. 2020. 'Countering the Conceits of the Anthropos: Scaling Down and Researching with Minor Players'. *Discourse: Studies in the Cultural Politics of Education* 41, no. 3: 340–58.

Taylor, Diane. 2022. 'Ukrainian Refugees Could Face Months of Separation from Pets Under UK Rules'. *The Guardian*, 25 March. https://www.theguardian.com/world/2022/mar/25/ukrainian-refugees-separation-pets-under-uk-rules-quarantine#:~:text=It%20is%20estimated%20that%20over,travel%20specialists%20PBS%20Pet%20Travel.

Taylor, Nik. 2011. 'Can Sociology Contribute to the Emancipation of Animals?'. In *Theorizing Animals. Re-thinking Humanimal Relations*, edited by Nik Taylor and Tania Signal, 203–20. Leiden: Brill.

Taylor, Nik. 2012. 'Animals, Mess, Method: Post-Humanism, Sociology and Animal Studies'. In *Crossing Boundaries. Investigating Human-Animal Relationships* edited by Lynda Birke and Jo Hockenhull, 37–50. Leiden: Brill.

Taylor, Nik, and Heather Fraser. 2018. 'Resisting Sexism and Speciesism in the Social Sciences: Using Feminist, Species-Inclusive, Visual Methods to Value the Work of Women and (Other) Animals'. *Gender, Work and Organization* 26, no. 3: 343–57.

Taylor, Nik, and Heather Fraser. 2019. *Companion Animals and Domestic Violence: Rescuing Me, Rescuing You*. Basingstoke: Palgrave.

Taylor, Nik, and Richard Twine. 2014. 'Introduction'. In *From the Margins to the Centre? The Rise of Critical Animal Studies*, edited by Nik Taylor and Richard Twine, 1–16. Abingdon: Routledge.

Taylor, Sunaura. 2017. *Beasts of Burden: Animal and Disability Liberation*. New York: New Press.

Tester, Keith. 1992. *Animals and Society: The Humanity of Animal Rights*. London: Routledge.

Timeto, Federica. 2021. 'Becoming-with in a Compost Society: Haraway beyond Posthumanism'. *International Journal of Sociology and Social Policy* 41, no. 3/4: 315–30.

Tipper, Becky. 2011. '"A Dog that I Know Quite Well": Everyday Relationships Between Children and Animals'. *Children's Geographies* 9, no. 2: 145–65.

The Scotsman. 2017. Woman Wins Right to Legally Use Sick Leave to Care for Dog, *The Scotsman*, 12 October. Accessed 4 November 2022. https://www.scotsman.com/news/woman-wins-right-legally-use-sick-leave-care-dog-1438043.

The Times. 2014. 'Lassie Comes Home to Save the Children One More Time'. *The Times*, 18 August. https://www.thetimes.co.uk/article/lassie-comes-home-to-save-the-children-one-more-time-3knblt63bqm.

Thorpe, Ronald J., Eleanor M. Simonsick, Jennifer S. Brach, Hilsa Anonayon, Suzanne Satterfield, Tamara B. Harris, Melissa Garcia and Stephen B. Kritchevsky. 2006. 'Dog Ownership, Walking Behaviour and Maintained Mobility in Late Life'. *Journal of American Geriatric Society* 54, no. 9: 1419–24.

Torres, Bob. 2007. *Making a Killing: The Political Economy of Animal Rights*. Oakland, CA: AK Press.

Tronto, Joan. 1993. *Moral Boundaries: A Political Argument for an Ethic of Care*. London: Psychology Press.

Tronto, Joan. 2013. *Caring Democracy: Markets, Equality and Justice*. New York: New York University Press.

Tsing, Anna. 2011. 'Arts of Inclusion, or, How to Love a Mushroom'. *Australian Humanities Review* 50, no. 1: 5–22.

Tsing, Anna. 2015. *The Mushroom at the End of the World: On the Possibility of Life in Capitalist Ruins*. Princeton, NJ: Princeton University Press.

Tuan, Yi-Fu. 1977. *Space and Place: The Perspective of Experience*. Minneapolis: University of Minnesota Press.

Tuan, Yi-Fu. 1984. *Dominance and Affection: The Making of Pets*. New Haven, CT: Yale University Press.

Tuan, Yi-Fu. 1986. 'Your Dog Is Sizing You Up'. *New York Times*, 7 September, section 7, p. 10.

Tuan, Yi-Fu. 1990. *Topophilia: A Study of Environmental Perceptions and Values*. New York: Colombia University Press, Morningside Edition.

Tuan, Yi-Fu. 1996. *Cosmos and Hearth: A Cosmopolitan Viewpoint*. Minneapolis: University of Minnesota Press.

Turner, Wendy G. 2001. 'Our New Children: The Surrogate Role of Companion Animals in Women's Lives'. *The Qualitative Report* 6, no. 1: 1–10.

Twine, Richard. 2012. 'Revealing the "Animal-Industrial Complex" – A Concept and Method for Critical Animal Studies'. *Journal for Critical Animal Studies* 10, no. 1: 12–39.

UK Pet Food. 2023 *Annual Report 2023: Fit for the Future*. Accessed 23 September 2023. https://ukpetfood-reports.co.uk/.

Valentine, Gill. 1993. '(Hetero)sexing Space: Lesbian Perceptions and Experiences of Everyday Spaces'. *Environment and Planning D: Society and Space* 11, no. 4: 395–413.

Valentine, Gill. 1996. 'Children Should be Seen and Not Heard: The Production and Transgression of Adult's Public Space'. *Urban Geography* 17, no. 3: 205–20.

Valentine, Gill. 1999. '"Oh Please Mum, Oh Please Dad": Negotiating Children's Spatial Boundaries'. In *Gender, Power and the Household*, edited by Linda McKie, Sophia Bowlby, Susan Gregory and Jo Campling, 137–54. Basingstoke: Macmillan.

van Dooren, Thom, and Deborah Bird Rose. 2016. 'Lively Ethnography Storying Animist Worlds'. *Environmental Humanities* 8, no. 1: 77–94.

van Dooren, Thom, Eben Kirksey and Ursula Münster. 2016. 'Multispecies Studies Cultivating Arts of Attentiveness'. *Environmental Humanities* 8, no. 1: 1–23.

Varner, Gary. 2002. 'Pets, Companion Animals and Domesticated Partners'. In *Ethics for Everyday*, edited by David Benatar, 450–75. New York: McGraw Hill.

Vegepets. n.d. 'About Us'. https://www.veggiepets.com/.

Veissier, Isabelle, and Mara Miele. 2014. 'Animal Welfare: Towards Transdisciplinarity – The European Experience'. *Animal Production Science* 54, no. 9: 1119–29.

Véron, Ophélie. 2016. '(Extra)ordinary Activism: Veganism and the Shaping of Hemerotopias'. *International Journal of Sociology and Social Policy* 36, no. 11/12: 756–73

Vetlesen, Arne J. 2019. *Cosmologies of the Anthropocene: Pansyschism, Animism, and the Limits of Posthumanism*. London: Routledge.

Vidler, Anthony. 2000. *Warped Space: Art, Architecture and Anxiety in Modern Culture*. Cambridge, MA: MIT Press.

Viveiros de Castro, Eduardo. 1998. 'Cosmological Deixis and Amerindian Perspectivism'. *Journal of the Royal Anthropological Institute* 4, no. 3: 469–88.

Volsche, Shelly, and Peter B. Gray. 2016. '"Dog Moms" Use Authoritative Parenting Styles'. *Human Animal Interact Bulletin* 4, no. 2: 1–16.

Volsche, Shelly. 2018. 'Negotiated Bonds: The Practice of Child-free Pet Parenting'. *Anthrozoos* 31, no. 3: 367–77.

Volsche, Shelly. 2019. 'Understanding Cross Species Parenting: A Case for Pets as Children'. In *Clinician's Guide to Treating Companion Animal Issues Addressing Human-Animal Interaction*, edited by Lori R. Kogan and Christopher Blazina, 129–41. London: Academic Press.

Wadiwell, Dinesh. 2020. 'The Working Day: Animals, Capitalism and Surplus Time'. In *Animal Labour: A New Frontier of Interspecies Justice?* edited by Charlotte E. Blattner, Kendra Coulter and Will Kymlica, 181–206. Oxford: Oxford University Press.

Walsh, Froma. 2009. 'Human-Animal Bonds II: The Role of Pets in Family Systems and Family Therapy. *Family Process* 48, no. 4: 481–99.

Weaver, Harlan. 2021. *Bad Dog: Pit Bull Politics and Multispecies Justice*. Seattle: University of Washington Press.

Wedderburn, Peter. 2010. 'Dangerous Dogs: Time for a New Approach'. *Daily Telegraph*, 5 July. http://blogs.telegraph.co.uk/news/peterwedderburn.

Weeks, Jeffrey, Catherine Donovan and Brain Heaphy. 2001. *Same Sex Intimacies: Families of Choice and Other Life Experiments*. London: Routledge.

Weeks, Kathi. 2011. *The Problem with Work: Feminism, Marxism, Antiwork Politics and Postwork Imaginaries*. Durham, NC: Duke University Press.

Weisberg, Zipporah. 2009. 'The Broken Promises of Monsters: Haraway, Animals and the Humanist Legacy'. *Journal for Critical Animal Studies* 7, no. 2: 22–62.

Wells, Deborah L. 2004. 'The Facilitation of Social Interactions by Domestic Dogs'. *Anthrozoös* 17, no. 4: 340–52.

Wells, Deborah L. 2007. 'Domestic Dog and Human Health: An Overview'. *British Journal of Health Psychology* 12: 145–56.

Wells, Yvonne, and H. Rodi. 2008. 'Effects of Pet Ownership on the Health and Well-being of Older People'. *Australasian Journal of Ageing* 19, no. 3: 143–48. https://doi.org/10.1111/j.1741-6612.2000.tb00167.x.

Westerlaken, Michelle. 2020. 'Imagining Multispecies Worlds'. PhD dissertation, published by Malmo University, Sweden. https://mau.diva-portal.org/smash/record.jsf?pid=diva2%3A1435111&dswid=3267.

Westerlaken, Michelle. 2021. 'It Matters What Designs Design Designs: Speculations on Multispecies Worlding'. *Global Discourse* 11, nos. 1–2: 137–55.

Westgarth, Cari, Gina L. Pinchbeck, John W. S. Bradshaw, Susan Dawson, Rosalind M. Gaskell and Robert M. Christley. 2007. 'Factors Associated with Dog Ownership and Contact with Dogs in a UK Community'. *BMC Veterinary Research* 3, no. 5. http://www.biomedcentral.com/1746-6148/3/5.

Westgarth, Carri, Robert M. Christley, Gina L. Pinchbeck, Rosalind M. Gaskell, Susan Dawson and John W. S. Bradshaw. 2009. 'Walking the Dog: Exploration of the Contact Networks Between Dogs in a Community'. *Epidemiology and Infection* 137, no. 8: 1169–78.

Westgarth, Carri, Robert M. Christley, Gina L. Pinchbeck, Rosalind M. Gaskell, Susan Dawson and John W. S. Bradshaw. 2010. 'Dog Behaviour on Walks and the Effect of Use of the Leash'. *Applied Animal Behaviour Science* 125, no. 1: 38–46.

Weston, Kath. 1991. *Families We Choose: Lesbians, Gays, Kinship*. New York: Columbia Press.

Whatmore, Sarah. 2006. 'Materialist Returns: Practising Cultural Geographies in and for a More-than-Human World'. *Cultural Geographies* 13, no. 4: 600–610.

Wilkie, Rhoda. 2010. *Livestock/Deadstock: Working with Farm Animals from Birth to Slaughter*. Philadelphia: Temple University Press.

Wilkie, Rhoda. 2015a. 'Academic "Dirty Work": Mapping Scholarly Labor in a Tainted Mixed-Species Field'. Society & Animals 23, no. 3: 211–30.

Wilkie, Rhoda. 2015b. 'Multispecies Scholarship and Encounters: Changing Assumptions at the Human-Animal Nexus'. *Sociology* 49, no. 2: 323–39.

Williams, Erin, and Margo de Mello. 2007. *Why Animals Matter: The Case for Animal Protection*. Amherst, NY: Prometheus Books.

Wolfe, Cary. 2009. *What is Posthumanism?* Minneapolis: Minnesota University Press.

Wolfe, Cary. 2013. *Before the Law: Human and Other Animals in Biopolitical Frame*. Chicago and London: Chicago University Press.

Wood, Lisa J., Billie Giles-Corti, Max K. Bulsara and Darcy A. Bosch. 2007. 'More than a Furry Companion: The Ripple Effect of Companion Animals on Neighbourhood Interactions and Sense of Community'. *Society and Animals* 15, no. 1: 43–56.

Wright Mills, Charles. 1956. *The Power Elite*. New York: Oxford University Press.

Wright Mills, Charles. 1958. *The Causes of World War Three*. New York: Simon & Schuster.

Zamir, Shamoon. 2008. *The Cambridge Companion to W. E. B. du Bois*. Cambridge: Cambridge University Press.

Zamir, Tzachi. 2006. 'The Moral Basis of Animal-Assisted Therapy'. *Society and Animals* 14, no. 2: 179–99.

Index

Page references for figures are italicized.

About the Author

Erika Cudworth works in the School of Applied Social Sciences at De Montfort University, Leicester, UK. Erika thinks of herself primarily as a sociologist, but her work also engages with politics and international relations, cultural studies and geography. Her research interests coalesce around complexity and posthumanist approaches, intersectionality and asking feminist questions of our relations with nonhuman animals and the earth. Her previous books include *Developing Ecofeminist Theory* and *Social Lives with Other Animals*, and, with Steve Hobden, *Posthuman International Relations* and *The Emancipatory Project of Posthumanism*. She has edited many collections, most recently, with Ruth McKie and Di Turgoose, *Feminist Animal Studies*. Erika is a Londoner who finds herself living in a Leicestershire village where walking with dogs reminds her, on a daily basis, that there are some animals who are much loved, and that there are many who are killed and eaten.

www.ingramcontent.com/pod-product-compliance
Ingram Content Group UK Ltd.
Pitfield, Milton Keynes, MK11 3LW, UK
UKHW041835031026
15587UKWH00002B/786

9 781538 180198